The Exploration of the Colorado River and Its Canyons

BY

John Wesley Powell

The Exploration of the Colorado River and Its Canyons

www.simonandbrown.com

The Exploration of the Colorado River and Its Canyons

John Wesley Powell

Formerly Director of the United States Geological Survey. Member of the National Academy of Sciences, etc., etc.

The Exploration of the Colorado River and Its Canyons

Contents

PREFACE.

On my return from the first exploration of the canyons of the Colorado, I found that our journey had been the theme of much newspaper writing. A story of disaster had been circulated, with many particulars of hardship and tragedy, so that it was currently believed throughout the United States that all the members of the party were lost save one. A good friend of mine had gathered a great number of obituary notices, and it was interesting and rather flattering to me to discover the high esteem in which I had been held by the people of the United States. In my supposed death I had attained to a glory which I fear my continued life has not fully vindicated.

The exploration was not made for adventure, but purely for scientific purposes, geographic and geologic, and I had no intention of writing an account of it, but only of recording the scientific results. Immediately on my return I was interviewed a number of times, and these interviews were published in the daily press; and here I supposed all interest in the exploration ended. But in 1874 the editors of Scribner's Monthly requested me to publish a popular account of the Colorado exploration in that journal. To this I acceded and prepared four short articles, which were elaborately illustrated from photographs in my possession.

In the same year–1874–at the instance of Professor Henry of the Smithsonian Institution, I was called before an appropriations committee of the House of Representatives to explain certain estimates made by the Professor for funds to continue scientific work which had been in progress from the date of the original exploration. Mr. Garfield was chairman of the committee, and after listening to my account of the progress of the geographic and geologic work, he asked me why no history of the original exploration of the canyons had been published. I informed him that I had no interest in that work as an adventure, but was interested only in the scientific results, and that these results had in part been published and in part were in course of publication. Thereupon Mr. Garfield, in a pleasant manner, insisted that the history of the exploration should be published by the government, and that I must understand that my scientific work would be continued by additional appropriations only upon my promise that I would publish an account of the exploration. I made the promise, and the task was immediately undertaken.

My daily journal had been kept on long and narrow strips of brown paper, which were gathered into little volumes that were bound in sole leather in camp as they were completed. After some deliberation I decided to publish this journal, with only such emendations and corrections as its hasty writing in camp necessitated. It chanced that the journal was written in the present tense, so that the first account of my trip appeared in that tense. The journal thus published was not a lengthy paper, constituting but a part of a report entitled "Exploration of the Colorado River of the West and its Tributaries. Explored in 1869, 1870, 1871, and 1872, under the direction of the Secretary of the Smithsonian Institution." The other papers published with it relate to the geography, geology, and natural history of the country. And here again I supposed all account of the exploration ended. But from that time until the present I have received many letters urging that a popular account of the exploration and a description of that wonderful land should be published by me. This call has been voiced occasionally in the daily press and sometimes in the magazines, until at last I have concluded to publish a fuller account in popular form. In doing this I have revised and enlarged the original journal of exploration, and have added several new chapters descriptive of the region and of the people who inhabit it. Realizing the difficulty of painting in word colors a land so strange, so wonderful, and so vast in its features, in the weakness of my descriptive powers I have sought refuge in graphic illustration, and for this purpose have gathered from the magazines and from various scientific reports an abundance of material. All of this illustrative material originated in my work, but it has already been used elsewhere.

Many years have passed since the exploration, and those who were boys with me in the enterprise are–ah, most of them are dead, and the living are gray with age. Their bronzed, hardy, brave faces come before me as they appeared in the vigor of life; their lithe but powerful forms seem to move around me; and the memory of the men and their heroic deeds, the men and their generous acts, overwhelms me with a joy that seems almost a grief, for it starts a fountain of tears. I was a maimed man; my right arm was gone; and these brave men, these good men, never forgot it. In every danger my safety was their first care, and in every waking hour some kind service was rendered me, and they transfigured my misfortune into a boon.

To you–J. C. Sumner, William H. Dunn, W. H. Powell, G. Y. Bradley, O. G. Howland, Seneca Howland, Prank Goodman, W. E. Hawkins, and Andrew Hall–my noble and generous companions, dead and alive, I dedicate this book.

CHAPTER I. THE VALLEY OF THE COLORADO.

The Colorado River is formed by the junction of the Grand and Green.

The Grand River has its source in the Rocky Mountains, five or six miles west of Long's Peak. A group of little alpine lakes, that receive their waters directly from perpetual snowbanks, discharge into a common reservoir known as Grand Lake, a beautiful sheet of water. Its quiet surface reflects towering cliffs and crags of granite on its eastern shore, and stately pines and firs stand on its western margin.

The Green River heads near Fremont's Peak, in the Wind River Mountains. This river, like the Grand, has its sources in alpine lakes fed by everlasting snows. Thousands of these little lakes, with deep, cold, emerald waters, are embosomed among the crags of the Rocky Mountains. These streams, born in the cold, gloomy solitudes of the upper mountain region, have a strange, eventful history as they pass down through gorges, tumbling in cascades and cataracts, until they reach the hot, arid plains of the Lower Colorado, where the waters that were so clear above empty as turbid floods into the Gulf of California.

The mouth of the Colorado is in latitude 31 degrees 53 minutes and longitude 115 degrees. The source of the Grand River is in latitude 40 degrees 17' and longitude 105 degrees 43' approximately. The source of the Green River is in latitude 43 degrees 15' and longitude 109 degrees 54' approximately.

The Green River is larger than the Grand and is the upper continuation of the Colorado. Including this river, the whole length of the stream is about 2,000 miles. The region of country drained by the Colorado and its tributaries is about 800 miles in length and varies from 300 to 500 miles in width, containing about 300,000 square miles, an area larger than all the New England and Middle States with Maryland, Virginia and West Virginia added, or nearly as large as Minnesota, Wisconsin, Iowa, Illinois, and Missouri combined.

There are two distinct portions of the basin of the Colorado, a desert portion below and a plateau portion above. The lower third, or desert portion of the basin, is but little above the level of the sea, though here and there ranges of mountains rise to an altitude of from 2,000 to 6,000 feet. This part of the valley is bounded on the northeast by a line of cliffs, which present a bold, often vertical step, hundreds or thousands of feet to the table-lands above. On the California side a vast desert stretches westward, past the head of the Gulf of California, nearly to the shore of the Pacific. Between the desert and the sea a narrow belt of valley, hill, and mountain of wonderful beauty is found. Over this coastal zone there falls a balm distilled from the great ocean, as gentle showers and refreshing dews bathe the land. When rains come the emerald hills laugh with delight as bourgeoning bloom is spread in the sunlight. When the rains have ceased all the verdure turns to gold. Then slowly the hills are brinded until the rains come again, when verdure and bloom again peer through the tawny wreck of the last year's greenery. North of the Gulf of California the desert is known as "Coahuila Valley," the most desolate region on the continent. At one time in the geologic history of this country the Gulf of California extended a long distance farther to the northwest, above the point where the Colorado River now enters it; but this stream brought its mud from the mountains and the hills above and poured it into the gulf and gradually erected a vast dam across it, until the waters above were separated from the waters below; then the Colorado cut a channel into the lower gulf. The upper waters, being cut off from the sea, gradually evaporated, and what is known as Coahuila Valley was the bottom of this ancient upper gulf, and thus the land is now below the level of the sea. Between Coahuila Valley and the river there are many low, ashen-gray mountains standing in short ranges. The rainfall is so little that no perennial streams are formed. When a great rain comes it washes the mountain sides and gathers on its way a deluge of sand, which it spreads over the plain below, for the streams do not carry the sediment to the sea. So the mountains are washed down and the valleys are filled. On the Arizona side of the river desert plains are interrupted by desert mountains. Far to the eastward the country rises until the Sierra Madre are reached in

New Mexico, where these mountains divide the waters of the Colorado from the Rio Grande del Norte. Here in New Mexico the Gila River has its source. Some of its tributaries rise in the mountains to the south, in the territory belonging to the republic of Mexico, but the Gila gathers the greater part of its waters from a great plateau on the northeast. Its sources are everywhere in pine-clad mountains and plateaus, but all of the affluents quickly descend into the desert valley below, through which the Gila winds its way westward to the Colorado. In times of continued drought the bed of the Gila is dry, but the region is subject to great and violent storms, and floods roll down from the heights with marvelous precipitation, carrying devastation on their way. Where the Colorado River forms the boundary between California and Arizona it cuts through a number of volcanic rocks by black, yawning canyons. Between these canyons the river has a low but rather narrow flood plain, with cottonwood groves scattered here and there, and a chaparral of mesquite bearing beans and thorns. Four hundred miles above its mouth and more than two hundred miles above the Gila, the Colorado has a second tributary–"Bill Williams' River" it is called by excessive courtesy. It is but a muddy creek. Two hundred miles above this the Rio Virgen joins the Colorado. This river heads in the Markagunt Plateau and the Pine Valley Mountains of Utah. Its sources are 7,000 or 8,000 feet above the sea, but from the beautiful course of the upper region it soon drops into a great sandy valley below and becomes a river of flowing sand. At ordinary stages it is very wide but very shallow, rippling over the quicksands in tawny waves. On its way it cuts through the Beaver Mountains by a weird canyon. On either side grease-wood plains stretch far away, interrupted here and there by bad-land hills.

The region of country lying on either side of the Colorado for six hundred miles of its course above the gulf, stretching to Coahuila Valley below on the west and to the highlands where the Gila heads on the east, is one of singular characteristics. The plains and valleys are low, arid, hot, and naked, and the volcanic mountains scattered here and there are lone and desolate. During the long months the sun pours its heat upon the rocks and sands, untempered by clouds above or forest shades beneath. The springs are so few in number that their names are household words in every Indian rancheria and every settler's home; and there are no brooks, no creeks, and no rivers but the trunk of the Colorado and the trunk of the Gila. The few plants are strangers to the dwellers in the temperate zone. On the mountains a few junipers and pinons are found, and cactuses, agave, and yuccas, low, fleshy plants with bayonets and thorns. The landscape of vegetal life is weird–no forests, no meadows, no green hills, no foliage, but clublike stems of plants armed with stilettos. Many of the plants bear gorgeous flowers. The birds are few, but often of rich plumage. Hooded rattlesnakes, horned toads, and lizards crawl in the dust and among the rocks. One of these lizards, the "Gila monster," is poisonous. Rarely antelopes are seen, but wolves, rabbits, and sundry ground squirrels abound.

The desert valley of the Colorado, which has been described as distinct from the plateau region above, is the home of many Indian tribes. Away up at the sources of the Gila, where the pines and cedars stand and where creeks and valleys are found, is a part of the Apache land. These tribes extend far south into the republic of Mexico. The Apaches are intruders in this country, having at some time, perhaps many centuries ago, migrated from British America. They speak an Athapascan language. The Apaches and Navajos are the American Bedouins. On their way from the far North they left several colonies in Washington, Oregon, and California. They came to the country on foot, but since the Spanish invasion they have become skilled horsemen. They are wily warriors and implacable enemies, feared by all other tribes. They are hunters, warriors, and priests, these professions not yet being differentiated. The cliffs of the region have many caves, in which these people perform their religious rites. The Sierra Madre formerly supported abundant game, and the little Sonora deer was common. Bears and mountain lions were once found in great numbers, and they put the courage and prowess of the Apaches to a severe test. Huge rattlesnakes are common, and the rattlesnake god is one of the deities of the tribes.

In the valley of the Gila and on its tributaries from the northeast are the Pimas, Maricopas, and Papagos. They are skilled agriculturists, cultivating lands by irrigation. In the same region many ruined villages are found. The dwellings of these towns in the valley were built chiefly of grout, and the fragments of the ancient pueblos still remaining have stood through centuries of storm. Other pueblos near the cliffs on the northeast were built of stone. The people who occupied them cultivated the soil by irrigation, and their hydraulic works were on an extensive scale. They built canals scores of miles in length and built reservoirs to store water. They were skilled workers in pottery. From the fibers of some of the desert plants they made fabrics with which to clothe themselves, and they cultivated cotton. They were deft artists in picture-writings, which they etched on the rocks. Many interesting vestiges of their ancient art remain, testifying to their skill as savage artisans. It seems probable that the Pimas, Maricopas, and Papagos are the same people who built the pueblos and constructed the irrigation works; so their traditions state. It is also handed down that the pueblos were destroyed in wars with the Apaches. In these groves of the flood plain of the Colorado the Mojave and Yuma Indians once had their homes. They caught fish from the river and snared a few rabbits in the desert, but lived mainly on mesquite beans, the hearts of yucca plants, and the fruits of the cactus. They also gathered a harvest from the river reeds. To some slight extent they cultivated the soil by rude irrigation and raised corn and squashes. They lived almost naked, for the climate is warm and dry. Sometimes a year passes without a drop of rain. Still farther to the north the Chemehuevas lived, partly along the river and partly in the mountains to the west, where a few springs are found. They belong to the great Shoshonian family. On the Rio Virgen and in the mountains round about, a confederacy of tribes speaking the Ute language and belonging to the Shoshonian family have their homes. These people built their sheltering homes of boughs and the bast of the juniper. In such shelters, they lived in winter, but in summer they erected extensive booths of poles and willows, sometimes large enough for the accommodation of a tribe of 100 or 200 persons. A wide gap in culture separates the Pimas, Maricopas, and Papagos from the Chemehuevas. The first were among the most advanced tribes found in the United States; the last were among the very lowest; they are the original "Digger" Indians, called so by all the other tribes, but the name has gradually spread beyond its original denotation to many tribes of Utah, Nevada, and California.

The low desert, with its desolate mountains, which has thus been described is plainly separated from the upper region of plateau by the Mogollon Escarpment, which, beginning in the Sierra Madre of New Mexico, extends northwestward across the Colorado far into Utah, where it ends on the margin of the Great Basin. The rise by this escarpment varies from 3,000 to more than 4,000 feet. The step from the lowlands to the highlands which is here called the Mogollon Escarpment is not a simple line of cliffs, but is a complicated and irregular facade presented to the southwest. Its different portions have been named by the people living below as distinct mountains, as Shiwits Mountains, Mogollon Mountains, Pinal Mountains, Sierra Calitro, etc., but they all rise to the summit of the same great plateau region.

The upper region, extending to the headwaters of the Grand and Green Rivers, constitutes the great Plateau Province. These plateaus are drained by the Colorado River and its tributaries; the eastern and southern margin by the Rio Grande and its tributaries, and the western by streams that flow into the Great Basin and are lost in the Great Salt Lake and other bodies of water that have no drainage to the sea. The general surface of this upper region is from 5,000 to 8,000 feet above sea level, though the channels of the streams are cut much lower.

This high region, on the east, north, and west, is set with ranges of snow-clad mountains attaining an altitude above the sea varying from 8,000 to 14,000 feet. All winter long snow falls on its mountain-crested rim, filling the gorges, half burying the forests, and covering the crags and peaks with a mantle woven by the winds from the waves of the sea. When the summer sun comes this snow melts and tumbles down the mountain sides in millions of cascades. A million cascade brooks unite to form a thousand torrent creeks; a thousand torrent creeks unite to form half a hundred rivers beset with cataracts; half a hundred roaring rivers unite to form the Colorado, which rolls, a mad, turbid stream, into the Gulf of California.

Consider the action of one of these streams. Its source is in the mountains, where the snows fall; its course, through the arid plains. Now, if at the river's flood storms were falling on the plains, its channel would be cut but little faster than the adjacent country would be washed, and the general level would thus be preserved; but under the conditions here mentioned, the river continually deepens its beds; so all the streams cut deeper and still deeper, until their banks are towering cliffs of solid rock. These deep, narrow gorges are called canyons.

For more than a thousand miles along its course the Colorado has cut for itself such a canyon; but at some few points where lateral streams join it the canyon is broken, and these narrow, transverse valleys divide it into a series of canyons.

The Virgen, Kanab, Paria, Escalante, Fremont, San Rafael, Price, and Uinta on the west, the Grand, White, Yampa, San Juan, and Colorado Chiquito on the east, have also cut for themselves such narrow winding gorges, or deep canyons. Every river entering these has cut another canyon; every lateral creek has cut a canyon; every brook runs in a canyon; every rill born of a shower and born again of a shower and living only during these showers has cut for itself a canyon; so that the whole upper portion of the basin of the Colorado is traversed by a labyrinth of these deep gorges.

Owing to a great variety of geological conditions, these canyons differ much in general aspect. The Rio Virgen, between Long Valley and the Mormon town of Rockville, runs through Parunuweap Canyon, which is often not more than 20 or 30 feet in width and is from 600 to 1,500 feet deep. Away to the north the Yampa empties into the Green by a canyon that I essayed to cross in the fall of 1868, but was baffled from day to day, and the fourth day had nearly passed before I could find my way down to the river. But thirty miles above its mouth this canyon ends, and a narrow valley with a flood plain is found. Still farther up the stream the river comes down through another canyon, and beyond that a narrow valley is found, and its upper course is now through a canyon and now through a valley. All these canyons are alike changeable in their topographic characteristics.

The longest canyon through which the Colorado runs is that between the mouth of the Colorado Chiquito and the Grand Wash, a distance of 217 1/2 miles. But this is separated from another above, 65 1/2 miles in length, only by the narrow canyon valley of the Colorado Chiquito.

All the scenic features of this canyon land are on a giant scale, strange and weird. The streams run at depths almost inaccessible, lashing the rocks which beset their channels, rolling in rapids and plunging in falls, and making a wild music which but adds to the gloom of the solitude. The little valleys nestling along the streams are diversified by bordering willows, clumps of box elder, and small groves of cottonwood.

Low mesas, dry, treeless, stretch back from the brink of the canyon, often showing smooth surfaces of naked, solid rock. In some places the country rock is composed of marls, and here the surface is a bed of loose, disintegrated material through which one walks as in a bed of ashes. Often these marls are richly colored and variegated. In other places the country rock is a loose sandstone, the disintegration of which has left broad stretches of drifting sand, white, golden, and vermilion. Where this sandstone is a conglomerate, a paving of pebbles has been left,–a mosaic of many colors, polished by the drifting sands and glistening in the sunlight.

After the canyons, the most remarkable features of the country are the long lines of cliffs. These are bold escarpments scores or hundreds of miles in length,–great geographic steps, often hundreds or thousands of feet in altitude, presenting steep faces of rock, often vertical. Having climbed one of these steps, you may descend by a gentle, sometimes imperceptible, slope to the foot of another. They thus present a series of terraces, the steps of which are well-defined escarpments of rock. The lateral extension of such a line of cliffs is usually very irregular; sharp salients are projected on the plains below, and deep recesses are cut into the terraces above. Intermittent streams coming down the cliffs have cut many canyons or canyon valleys, by which the traveler may pass from the plain below to the terrace above. By these gigantic stairways he may ascend to high plateaus, covered with forests of pine and fir.

The region is further diversified by short ranges of eruptive mountains. A vast system of fissures–huge cracks in the rocks to the depths below–extends across the country. From these crevices floods of lava have poured, covering mesas and table-lands with sheets of black basalt. The expiring energies of these volcanic agencies have piled up huge cinder cones that stand along the fissures, red, brown, and black, naked of vegetation, and conspicuous landmarks, set as they are in contrast to the bright, variegated rocks of sedimentary origin.

These canyon gorges, obstructing cliffs, and desert wastes have prevented the traveler from penetrating the country, so that until the Colorado River Exploring Expedition was organized it was almost unknown. In the early history of the country Spanish adventurers penetrated the region and told marvelous stories of its wonders. It was also traversed by priests who sought to convert the Indian tribes to Christianity. In later days, since the region has been under the control of the United States, various government expeditions have penetrated the land. Yet enough had been seen in the earlier days to foment rumor, and many wonderful stories were told in the hunter's cabin and the prospector's camp–stories of parties entering the gorge in boats and being carried down with fearful velocity into whirlpools where all were overwhelmed in the abyss of waters, and stories of underground passages for the great river into which boats had passed never to be seen again. It was currently believed that the river was lost under the rocks for several hundred miles. There were other accounts of great falls whose roaring music could be heard on the distant mountain summits; and there were stories current of parties wandering on the brink of the canyon and vainly endeavoring to reach the waters below, and perishing with thirst at last in sight of the river which was roaring its mockery into their dying ears.

The Indians, too, have woven the mysteries of the canyons into the myths of their religion. Long ago there was a great and wise chief who mourned the death of his wife and would not be comforted, until Tavwoats, one of the Indian gods, came to him and told him his wife was in a happier land, and offered to take him there that he might see for himself, if, upon his return, he would cease to mourn. The great chief promised. Then Tavwoats made a trail through the mountains that intervene between that beautiful land, the balmy region of the great west, and this, the desert home of the poor Numa. This trail was the canyon gorge of the Colorado. Through it he led him; and when they had returned the deity exacted from the chief a promise that he would tell no one of the trail. Then he rolled a river into the gorge, a mad, raging stream, that should engulf any that might attempt to enter thereby.

CHAPTER II. MESAS AND BUTTES.

From the Grand Canyon of the Colorado a great plateau extends southeastward through Arizona nearly to the line of New Mexico, where this elevated land merges into the Sierra Madre. The general surface of this plateau is from 6,000 to 8,000 feet above the level of the sea. It is sharply defined from the lowlands of Arizona by the Mogollon Escarpment. On the northeast it gradually falls off into the valley of the Little Colorado, and on the north it terminates abruptly in the Grand Canyon.

Various tributaries of the Gila have their sources in this escarpment, and before entering the desolate valley below they run in beautiful canyons which they have carved for themselves in the margin of the plateau. Sometimes these canyons are in the sandstones and limestones which constitute the platform of the great elevated region called the San Francisco Plateau. The escarpment is caused by a fault, the great block of the upper side being lifted several thousand feet above the valley region. Through the fissure lavas poured out, and in many places the escarpment is concealed by sheets of lava. The canyons in these lava beds are often of great interest.

On the plateau a number of volcanic mountains are found, and black cinder cones are scattered in profusion. Through the forest lands are many beautiful prairies and glades that in midsummer are decked with gorgeous wild flowers. The rains of the region give source to few perennial streams, but intermittent streams have carved deep gorges in the plateau, so that it is divided into many blocks. The upper surface, although forest-clad and covered with beautiful grasses, is almost destitute of water. A few springs are found, but they are far apart, and some of the volcanic craters hold lakelets. The limestone and basaltic rocks sometimes hold pools of water; and where the basins are deep the waters are perennial. Such pools are known as "water pockets."

This is the great timber region of Arizona. Not many years ago it was a vast park for elk, deer, and antelope, and bears and mountain lions were abundant. This is the last home of the wild turkey in the United States, for they are still found here in great numbers. San Francisco Peak is the highest of these volcanic mountains, and about it are grouped in an irregular way many volcanic cones, one of which presents some remarkable characteristics. A portion of the cone is of bright reddish cinders, while the adjacent rocks are of black basalt. The contrast in the colors is so great that on viewing the mountain from a distance the red cinders seem to be on fire. From this circumstance the cone has been named Sunset Peak. When distant from it ten or twenty miles it is hard to believe that the effect is produced by contrasting colors, for the peak seems to glow with a light of its own.

In centuries past the San Francisco Plateau was the home of pueblo-building tribes, and the ruins of their habitations are widely scattered over this elevated region. Thousands of little dwellings are found, usually built of blocks of basalt. In some cases they were clustered in little towns, and three of these deserve further mention.

A few miles south of San Francisco Peak there is an intermittent stream known as Walnut Creek. This stream runs in a deep gorge 600 to 800 feet below the general surface. The stream has cut its way through the limestone and through series of sandstones, and bold walls of rock are presented on either side. In some places the softer sandstones lying between the harder limestones and sandstones have yielded to weathering agencies, so that there are caves running along the face of the wall, sometimes for hundreds or thousands of feet, but not very deep. These natural shelves in the rock were utilized by an ancient tribe of Indians for their homes. They built stairways to the waters below and to the hunting grounds above, and lived in the caves. They walled the fronts of the caves with rock, which they covered with plaster, and divided them into compartments or rooms; and now many hundreds of these dwellings are found. Such is the cliff village of Walnut Canyon. In the ruins of these cliff houses mortars and pestles are found in great profusion, and when first discovered many articles of pottery were found, and still many potsherds are seen. The people were very skillful in the manufacture of stone implements, especially spears, knives, and arrows.

East of San Francisco Peak there is another low volcanic cone, composed of ashes which have been slightly cemented by the processes of time, but which can be worked with great ease. On this cone another tribe of Indians made its village, and for the purpose they sunk shafts into the easily worked but partially consolidated ashes, and after penetrating from the surface three or four feet they enlarged the chambers so as to make them ten or twelve feet in diameter. In such a chamber they made a little fireplace, its chimney running up on one side of the wellhole by which the chamber was entered. Often they excavated smaller chambers connected with the larger, so that sometimes two, three, four, or even five smaller connecting chambers are grouped about a large central room. The arts of these people resembled those of the people who dwelt in Walnut Canyon. One thing more is worthy of special notice. On the very top of the cone they cleared off a space for a courtyard, or assembly square, and about it they erected booths, and within the square a space of ground was prepared with a smooth floor, on which they performed the ceremonies of their religion and danced to the gods in prayer and praise.

Some twelve or fifteen miles farther east, in another volcanic cone, a rough crater is found, surrounded by piles of cinders and angular fragments of lava. In the walls of this crater many caves are found, and here again a village was established, the caves in the scoria being utilized as habitations of men. These little caves were fashioned into rooms of more symmetry and convenience than originally found, and the openings to the caves were walled. Nor did these people neglect the gods, for in this crater town, as in the cinder-cone town, a place of worship was prepared.

Many other caves opening into the canyon and craters of this plateau were utilized in like manner as homes for tribal people, and in one cave far to the south a fine collection of several hundred pieces of pottery has been made.

On the northeast of the San Francisco Plateau is the valley of the Little Colorado, a tributary of the Colorado River. This river is formed by streams that head chiefly on the San Francisco Plateau, but in part on the Zuni Plateau. The Little Colorado is a marvelous river. In seasons of great rains it is a broad but shallow torrent of mud; in seasons of drought it dwindles and sometimes entirely disappears along portions of its course. The upper tributaries usually run in beautiful box canyons. Then the river flows through a low, desolate, bad-land valley, and the river of mud is broad but shallow, except in seasons of great floods. But fifty miles or more above the junction of this stream with the Colorado River proper, it plunges into a canyon with limestone walls, and steadily this canyon increases in depth, until at the mouth of the stream it has walls more than 4,000 feet in height. The contrast between this canyon portion and the upper valley portion is very great. Above, the river ripples in a broad sheet of mud; below, it plunges with violence over great cataracts and rapids. Above, the bad lands stretch on either hand. This is the region of the Painted Desert, for the marls and soft rocks of which the hills are composed are of many colors–chocolate, red, vermilion, pink, buff, and gray; and the naked hills are carved in fantastic forms. Passing to the region below, suddenly the channel is narrowed and tumbles down into a deep, solemn gorge with towering limestone cliffs.

All round the margin of the valley of the Little Colorado, on the side next to the Zuni Plateau and on the side next to the San Francisco Plateau, every creek and every brook runs in a beautiful canyon. Then down in the valley there are stretches of desert covered with sage and grease wood. Still farther down we come to the bad lands of the Painted Desert; and scattered through the entire region low mesas or smaller plateaus are everywhere found.

On the northeast side of the Little Colorado a great mesa country stretches far to the northward. These mesas are but minor plateaus that are separated by canyons and canyon valleys, and sometimes by low sage plains. They rise from a few hundred to 2,000 or 3,000 feet above the lowlands on which they are founded. The distinction between plateaus and mesas is vague; in fact, in local usage the term mesa is usually applied to all of these tables which do not carry volcanic mountains. The mesas are carved out of platforms of horizontal or nearly horizontal rocks by perennial or intermittent streams, and as the climate is exceedingly arid most of the streams flow only during seasons of rain, and for the greater part of the year they are dry arroyos. Many of the longer channels are dry for long periods. Some of them are opened only by floods that come ten or twenty years apart.

The region is also characterized by many buttes. These are plateaus or mesas of still smaller dimensions in horizontal distance, though their altitude may be hundreds or thousands of feet. Like the mesas and plateaus, they sometimes form very conspicuous features of a landscape and are of marvelous beauty by reason of their sculptured escarpments. Below they are often buttressed on a magnificent scale. Softer beds give rise to a vertical structure of buttresses and columns, while the harder strata appear in great horizontal lines, suggesting architectural entablature. Then the strata of which these buttes are composed are of many vivid colors; so color and form unite in producing architectural effects, and the buttes often appear like Cyclopean temples.

There is yet one other peculiarity of this landscape deserving mention here. Before the present valleys and canyons were carved and the mesas lifted in relief, the region was one of great volcanic activity. In various places vents were formed and floods of lava poured in sheets over the land. Then for a time volcanic action ceased, and rains and rivers carved out the valleys and left the mesas and mountains standing. These same agencies carried away the lava beds that spread over the lands. But wherever there was a lava vent it was filled with molten matter, which on cooling was harder than the sandstones and marls through which it penetrated. The chimney to the region of fire below was thus filled with a black rock which yielded more slowly to the disintegrating agencies of weather, and so black rocks rise up from mesas on every hand. These are known as volcanic necks, and, being of a somber color, in great contrast with the vividly colored rocks from which they rise and by which they are surrounded, they lend a strange aspect to the landscape. Besides these necks, there are a few volcanic mountains that tower over all the landscape and gather about themselves the clouds of heaven. Mount Taylor, which stands over the divide on the drainage of the Rio Grande del Norte, is one of the most imposing of the dead volcanoes of this region. Still later eruptions of lava are found here and there, and in the present valleys and canyons sheets of black basalt are often found. These are known as coulees, and sometimes from these coulees cinder cones arise.

This valley of the Little Colorado is also the site of many ruins, and the villages or towns found in such profusion were of mueh larger size than those on the San Francisco Plateau. Some of the pueblo-building peoples yet remain. The Zuni Indians still occupy their homes, and they prove to be a most interesting people. They have cultivated the soil from time immemorial. They build their houses of stone and line them with plaster; and they have many interesting arts, being skilled potters and deft weavers. The seasons are about equally divided between labor, worship, and play.

A hundred miles to the northwest of the Zuni pueblo are the seven pueblos of Tusayan: Oraibi, Shumopavi, Shupaulovi, Mashongnavi, Sichumovi, Walpi, and llano. These towns are built on high cliffs. The people speak a language radically different from that of the Zuni, but, with the exception of that of the inhabitants of Hano, closely allied to that of the Utes. The people of Hano are Tewans, whose ancestors moved from the Rio Grande to Tusayan during the great Pueblo revolt against Spanish authority in 1680-96.

Between the Little Colorado and the Rio San Juan there is a vast system of plateaus, mesas, and buttes, volcanic mountains, volcanic cones, and volcanic cinder cones. Some of the plateaus are forest-clad and have perennial waters and are gemmed with lakelets. The mesas are sometimes treeless, but are often covered with low, straggling, gnarled cedars and pifions, trees that are intermediate in size between the bushes of sage in the desert and the forest trees of the elevated regions. On the western margin of this district the great Navajo Mountain stands, on the brink of Glen Canyon, and from its summit many of the stupendous gorges of the Colorado River can be seen. Central in the region stand the Carrizo Mountains, the Lukachukai Mountains, the Tunitcha Mountains, and the Chusca Mountains, which in fact constitute one system, extending from north to south in the order named. These are really plateaus crowned with volcanic peaks.

But the district we are now describing, which stretches from the Little Colorado to the San Juan, is best characterized by its canyons. The whole region is a labyrinth of gorges. On the west the Navajo Creek and its tributaries run in profound chasms. Farther south the Moencopie with its tributaries is a labyrinth of gorges; and all the streams that run west into the Colorado, south into the Little Colorado, or north into the San Juan have carved deep, wild, and romantic gorges. Immediately west of the Chusca Plateau the Canyon del Muerta and the Canyon de Chelly are especially noticeable. Many of these canyons are carved in a homogeneous red sandstone, and their walls are often vertical for hundreds of feet. Sometimes the canyons widen into narrow valleys, which are thus walled by impassable cliffs, except where lateral canyons cut their way through the battlements.

In these mountains, plateaus, mesas, and canyons the Navajo Indians have their home. The Navajos are intruders in this country. They belong to the Athapascan stock of British America and speak an Athapascan language, like the Apaches of the Sierra Madre country. They are a stately, athletic, and bold people. While yet this country was a part of Mexico they acquired great herds of horses and flocks of sheep, and lived in opulence compared with many of the other tribes of North America. After the acquisition of this territory by the United States they became disaffected by reason of encroaching civilization, and the petty wars between United States troops and the Navajos were in the main disastrous to our forces, due in part to the courage, skill, and superior numbers of the Navajos and in part to the character of the country, which is easily defended, as the routes of travel along the canyons present excellent opportunities for defense and ambuscade. But under the leadership and by the advice of Kit Carson these Indians were ultimately conquered. This wily but brave frontiersman recommended a new method of warfare, which was to destroy the herds and flocks of the Navajos; and this course was pursued. Regular troops with volunteers from California and New Mexico went into the Navajo country and shot down their herds of half-wild horses, killed hundreds of thousands of sheep, cut down their peach orchards which were scattered about the springs and little streams, destroyed their irrigating works, and devastated their little patches of corn, squashes, and melons; and entirely neglected the Navajos themselves, who were concealed among the rocks of the canyons. Seeing the destruction wrought upon their means of livelihood, the Navajos at once yielded. More than 8,000 of them surrendered at one time, coming in in straggling bands. They were then removed far to the east, near to the Texas line, and established on a reservation at the Bosque Redondo. Here they engaged in civilized farming. A great system of irrigation was developed; but the appropriations necessary for the maintenance of so large a body of people in the course of their passage from savagery to civilization seemed too great to those responsible for making grants from the national treasury, and just before 1870 the Navajos were permitted to break up their homes at the Bosque Redondo and return to the canyons and cliffs of their ancient land. Millions were spent in conquering them where thousands were used to civilize them, so that they were conquered but not civilized. Still, they are making good progress, and have once more acquired large flocks and herds. It is estimated that they now have more than a million sheep. Their experience in irrigation at the Bosque Redondo has not been wholly wasted, for they now cultivate the soil by methods of irrigation greatly improved over those used in the earlier time. Originally they dwelt in hogans, or houses made of poles arranged with much skill in conical form, the poles being covered with reeds and the reeds with earth; now they are copying the dwelling places of civilized men. They have also acquired great skill in the manufacture of silver ornaments, with which they decorate themselves and the trappings of their steeds.

Perhaps the most interesting ruins of America are found in this region. The ancient pueblos found here are of superior structure, but they were all built by a people whom the Navajos displaced when they migrated from the far North. Wherever there is water, near by an ancient ruin may be found; and these ruins are gathered about centers, the centers being larger pueblos and the scattered ruins representing single houses. The ancient people lived in villages, or pueblos, but during the growing season they scattered about by the springs and streams to cultivate the soil by irrigation, and wherever there was a little farm or garden patch, there was built a summer house of stone. When times of war came, especially when they were invaded by the Navajos, these ancient people left their homes in the pueblos and by the streams and constructed temporary homes in the cliffs and canyon walls. Such cliff ruins are abundant throughout the region, intimately the ancient pueblo peoples succumbed to the prowess of the Navajos and were driven out. A part joined related tribes in the valley of the Bio Grande; others joined the Zuni and the people of Tusayan; and stall others pushed on beyond the Little Colorado to the San Francisco Plateau and far down into the valley of the Gila.

Farther to the east, on the border of the region which we have described, beyond the drainage of the Little Colorado and San Juan and within the drainage of the Rio Grande, there lies an interesting plateau region, which forms a part of the Plateau Province and which is worthy of description. This is the great Tewan Plateau, which carries several groups of mountains. The western edge of this plateau is known as the Nacimiento Mountain, a long north-and-south range of granite, which presents a bold facade to the valley of the Puerco on the west. Ascending to the summit of this granite range, there is presented to the eastward a plateau of vast proportions, which stretches far toward Santa Fe and is terminated by the canyon of the Rio Grande del Norte. The eastern flank of this range as it slowly rose was a gentle slope, but as it came up fissures were formed and volcanoes burst forth and poured out their floods of lava, and now many extinct volcanoes can be seen. The plateau was built by these volcanoes–sheets of lava piled on sheets of lava hundreds and even thousands of feet in thickness. But with the floods of lava came great explosions, like that of Krakatoa, by which the heavens were filled with volcanic dust. These explosions came at different times and at different places, but they were of enormous magnitude, and when the dust fell again from the clouds it piled up in beds scores and hundreds of feet in thickness. So the Tewan Plateau has a foundation of red sandstone; upon this are piled sheets of lava and sheets of dust in many alternating layers. It is estimated that there still remain more than two hundred cubic miles of this dust, now compacted into somewhat coherent rocks and interpolated between sheets of lava. Everywhere this dust-formed rock is exceedingly light. Much of it has a specific gravity so low that it will float on water. Above the sheets of lava and above the beds of volcanic dust great volcanic cones rise, and the whole upper region is covered with forests interspersed with beautiful prairies. The plateau itself is intersected with many deep, narrow canyons, having walls of lava, volcanic dust, or tufa, and red sandstone. It is a beautiful region. The low mesas on every side are almost treeless and are everywhere deserts, but the great Tewan Plateau is booned with abundant rains, and it is thus a region of forests and meadows, divided into blocks by deep, precipitous canyons and crowned with cones that rise to an altitude of from 10,000 to 12,000 feet.

For many centuries the Tewan Plateau, with its canyons below and its meadows and forests above, has been the home of tribes of Tewan Indians, who built pueblos, sometimes of the red sandstones in the canyons, but oftener of blocks of tufa, or volcanic dust. This light material can be worked with great ease, and with crude tools of the harder lavas they cut out blocks of the tufa and with them built pueblos two or three stories high. The blocks are usually about twenty inches in length, eight inches in width, and six inches in thickness, though they vary somewhat in size. On the volcanic cones which dominate the country these people built shrines and worshiped their gods with offerings of meal and water and with prayer symbols made of the plumage of the birds of the air. When the Navajo invasion came, by which kindred tribes were displaced from the district farther west, these Tewan Indians left their pueblos on the plateau and their dwellings by the rivers below in the depths of the canyon and constructed cavate homes for themselves; that is, they excavated chambers in the cliffs where these cliffs were composed of soft, friable tufa. On the face of the cliff, hundreds of feet high and thousands of feet or even miles in length, they dug out chambers with stone tools, these chambers being little rooms eight or ten feet in diameter. Sometimes two or more such chambers connected. Then they constructed stairways in the soft rock, by which their cavate houses were reached; and in these rock shelters they lived during times of war. When the Navajo invasion was long past, civilized men as Spanish adventurers entered this country from Mexico, and again the Tewan peoples left their homes on the mesas and by the canyons to find safety in the cavate dwellings of the cliffs; and now the archaeologist in the study of this country discovers these two periods of construction and occupation of the cavate dwellings of the Tewan Indians.

North of the Rio San Juan another vast plateau region is found, stretching to the Grand River. The mountains of this region are the La Plata Mountains, Bear River Mountains, and San Miguel Mountains on the east, and the Sierra El Late, the Sierra Abajo, and the Sierra La Sal on the west, the latter standing near the brink of Cataract Canyon, through which the Colorado flows immediately below the junction of the Grand and Green. Throughout the region mountains, volcanic cones, volcanic necks, and coulees are found, while the mountains themselves rise to great altitudes and are forest-clad. Some of the plateaus attain huge proportions, and between the plateaus labyrinthian mesas are found. Buttes, as stupendous cameos, are scattered everywhere, and the whole region is carved with canyons.

Grand River heads on the back of Long's Peak, in the Front Range of the Rocky Mountains of central Colorado. At the foot of the mountain lies Grand Lake, a sheet of emerald water that duplicates the forest standing on its brink. Out of the lake flows Grand River, gathering on its way the many mountain streams whose waters fill the solitude with perennial music–a symphony of cascades. In Middle Park boiling springs issue from depths below and gather in pools covered with con-fervae. Leaving Middle Park the river goes through a great range known as the Gore's Pass Mountains; and still it flows on toward the Colorado, now through canyon and now through valley, until the last forty miles of its course it finds its way through a beautiful gorge known as Grand River Canyon. In its principal course this canyon is a bright red homogeneous sandstone, and the walls are often vertical and of great symmetry. Farther down, its walls are rugged and angular, being composed of limestones.

The principal tributaries from the south are the Blue, which heads in Mt. Lincoln, and the Gunnison, which heads in the Wasatch Mountains. These streams are also characterized by deep canyons and plateaus, and mesas abound on every hand. Between the Grand River and the White River, farther to the east, the Tavaputs Plateau is found. It begins at the foot of Gore's Pass Range and extends down between the rivers last mentioned to the very brink of Green River, which is in fact the upper Colorado. Between the Grand River and the foot of this plateau there is a low, narrow valley with mesas and buttes. Then the country suddenly rises by a stupendous line of cliffs 2,000 or 3,000 feet high. These cliffs are composed of sand stones, limestones, and shales, of many colors. The stratification in many places is minute, so that they have been called the Book Cliffs.

From the cliffs many salients are projected into the valleys, and within deep re-entering angles vast amphitheaters appear. About the projected salients many towering buttes, with pinnacles and minarets, are found. The long, narrow plateau is covered with a forest along its summit, and, though it rises abruptly on the south side from Grand River Valley, it descends more gently toward the White River, and on this slope many canyons of rare beauty are seen. Plateaus and mesas and canyons and buttes characterize the region north of White River and stretch out to the Yampa. The Yampa itself has an important tributary from the northwest, known as Snake River. Just below the affluence of the Snake with the Yampa a strange phenomenon is observed. Right athwart the course of the river rises a great dome-shaped mountain, with valley stretches on every side, and through this mountain the river runs, dividing it by a beautiful canyon, through which it flows to its junction with the Green. This canyon is in soft, white sandstone, usually with vertical walls varying from 500 to 2,000 feet in height, and the river flows in a gentle winding way through all this stretch. To the east of this plateau region, with its mesas and buttes and its volcanic mountains, stand the southern Rocky Mountains, or Park Mountains, a system of north-and-south ranges. These ranges are huge billows in the crust of the earth out of which mountains have been carved. The parks of Colorado are great valley basins enclosed by these ranges, and over their surfaces moss agates are scattered. The mountains are covered with dense forests and are rugged and wild. The higher peaks rise above the timber line and are naked gorges of rocks. In them the Platte and Arkansas rivers head and flow eastward to join the Missouri River. Here also heads the Rio Grande del Norte, which flows southward into the Gulf of Mexico, and still to the west head many streams which pour into the

Colorado waters destined for the Gulf of California. Throughout all of this region drained by the Grand, White, and Yampa rivers, there are many beautiful parks. The great mountain slopes are still covered with primeval forests. Springs, brooks, rivers, and lakes abound, and the waters are filled with trout. Not many years ago the hills were covered with game–elk on the mountains, deer on the plateaus, antelope in the valleys, and beavers building their cities on the streams. The plateaus are covered with low, dwarf oaks and many shrubs bearing berries, and in the chaparral of this region cinnamon bears are still abundant.

From time immemorial the region drained by the Grand, White, and Yampa rivers has been the home of Ute tribes of the Shoshonean family of Indians. These people built their shelters of boughs and bark, and to some extent lived in tents made of the skins of animals. They never cultivated the soil, but gathered wild seeds and roots and were famous hunters and fishermen. As the region abounds in game, these tribes have always been well clad in skins and furs. The men wore blouse, loincloth leggins, and moccasins, and the women dressed in short kilts. It is curious to notice the effect which the contact of civilization has had upon these women's dress. Even twenty years ago they had lengthened their skirts; and dresses, made of buckskin, fringed with furs, and beaded with elk teeth, were worn so long that they trailed on the ground. Neither men nor women wore any headdress except on festival occasions for decoration; then the women wore little basket bonnets decorated with feathers, and the men wore headdresses made of the skins of ducks, geese, eagles, and other large birds. Sometimes they would prepare the skin of the head of the elk or deer, or of a bear or mountain lion or wolf, for a headdress. For very cold weather both men and women were provided with togas for their protection. Sometimes the men would have a bearskin or elkskin for a toga; more often they made their togas by piecing together the skins of wolves, mountain lions, wolverines, wild cats, beavers, and otters. The women sometimes made theirs of fawnskins, but rabbitskin robes were far more common. These rabbitskins were tanned with the fur on, and cut into strips; then cords were made of the fiber of wild flax or yucca plants, and round these cords the strips of rabbitskin were rolled, so that they made long ropes of rabbitskin coils with a central cord of vegetal fiber; then these coils were woven in parallel strings with cross strands of fiber. The robe when finished was usually about five or six feet square, and it made a good toga for a cold day and a warm blanket for the night.

The Ute Indians, like all the Indians of North America, have a wealth of mythic stories. The heroes of these stories are the beasts, birds, and reptiles of the region, and the themes of the stories are the doings of these mythic beasts–the ancients from whom the present animals have descended and degenerated. The primeval animals were wonderful beings, as related in the lore of the Utes. They were the creators and controllers of all the phenomena of nature known to these simple-minded people. The Utes are zootheists. Each little tribe has its Shaman, or medicine man, who is historian, priest, and doctor. The lore of this Shaman is composed of mythic tales of ancient animals. The Indians are very skillful actors, and they represent the parts of beasts or reptiles, wearing masks and imitating the ancient zoic gods. In temples walled with gloom of night and illumed by torch fires the people gather about their Shaman, who tells and acts the stories of creation recorded in their traditional bible. When fever prostrates one of the tribe the Shaman gathers the actors about the stricken man, and with weird dancing, wild ululation, and ecstatic exhortation the evil spirit is driven from the body. Then they have their ceremonies to pray for the forest fruits, for abundant game, for successful hunting, and for prosperity in war.

CHAPTER III. MOUNTAINS AND PLATEAUS.

Green River has its source in Fremont's Peak, high up in the Wind River Mountains among glacial lakes and mountain cascades. This is the real source of the Colorado River, and it stands in strange contrast with the mouth of that stream where it pours into the Gulf of California. The general course of the river is from north to south and from great altitudes to the level of the sea. Thus it runs "from land of snow to land of sun." The Wind River Mountains constitute one of the most imposing ranges of the United States. Fremont's Peak, the culminating point, is 13,790 feet above the level of the sea. It stands in a wilderness of crags. Here at Fremont's Peak three great rivers have their sources: Wind River flows eastward into the Mississippi; Green River flows southward into the Colorado; and Gros Ventre River flows northwestward into the Columbia. From this dominating height many ranges can be seen on every hand. About the sources of the Platte and the Big Horn, that flow ultimately into the Gulf of Mexico, great ranges stand with their culminating peaks among the clouds; and the mountains that extend into Yellowstone Park, the land of geyser wonders, are seen. The Yellowstone Park is at the southern extremity of a great system of mountain ranges, the northern Rocky Mountains, sometimes called the Geyser Ranges. This geological province extends into British America, but its most wonderful scenery is in the upper Yellowstone basin, where geysers bombard the heavens with vapor distilled in subterranean depths. The springs which pour out their boiling waters are loaded with quartz, and the waters of the springs, flowing away over the rocks, slowly discharge their fluid magma, which crystallizes in beautiful forms and builds jeweled basins that hold pellucid waters.

To the north and west of Fremont's Peak are mountain ranges that give birth to rivers flowing into the great Columbia. Conspicuous among these from this point of view is the great Teton Range, with its towering facade of storm-carved rocks; then the Gros Ventre Mountains, the Snake River Range, the Wyoming Range, and, still beyond the latter, the Bear River Range, are seen. Far in the distant south, scarcely to be distinguished from the blue clouds on the horizon, stand the Uinta Mountains. On every hand are deep mountain gorges where snows accumulate to form glaciers. Below the glaciers throughout the entire Wind River Range great numbers of morainal lakes are found. These lakes are gems–deep sapphire waters fringed with emerald zones. From these lakes creeks and rivers flow, by cataracts and rapids, to form the Green. The mountain slopes below are covered with dense forests of pines and firs. The lakes are often fringed with beautiful aspens, and when the autumn winds come their golden leaves are carried over the landscape in clouds of resplendent sheen. The creeks descend from the mountains in wild rocky gorges, until they flow out into the valley. On the west side of the valley stand the Gros Ventre and the Wyoming mountains, low ranges of peaks, but picturesque in form and forest stretch. Leaving the mountain, the river meanders through the Green River Plains, a cold elevated district much like that of northern Norway, except that the humidity of Norway is replaced by the aridity of Wyoming. South of the plains the Big Sandy joins the Green from the east. South of the Big Sandy a long zone of sand-dunes stretches eastward. The western winds blowing up the valley drift these sands from hill to hill, so that the hills themselves are slowly journeying eastward on the wings of arid gales, and sand tempests may be encountered more terrible than storms of snow or hail. Here the northern boundary of the Plateau Province is found, for mesas and high table-lands are found on either side of the river.

On the east side of the Green, mesas and plateaus have irregular escarpments with points extending into the valleys, and between these points canyons come down that head in the highlands. Everywhere the escarpments are fringed with outlying buttes. Many portions of the region are characterized by bad lands. These are hills carved out of sandstone, shales, and easily disintegrated rocks, which present many fantastic forms and are highly colored in a great variety of tint and tone, and everywhere they are naked of vegetation. Now and then low mountains crown the plateaus. Altogether it is a region of desolation. Through the midst of the country, from east to west, flows an intermittent stream known as Bitter Creek. In seasons of rain it carries floods; in seasons of drought it disappears in the sands, and its waters are alkaline and often poisonous. Stretches of bad-land desert are interrupted by other stretches of sage plain, and on the high lands gnarled and picturesque forests of juniper and pinon are found. On the west side of the river the mesas rise by grassy slopes to the westward into high plateaus that are forest-clad, first with juniper and pinon, and still higher with pines and firs. Some of the streams run in canyons and others have elevated valleys along their courses. On the south border of this mesa and plateau country are the Bridger Bad Lands, lying at the foot of the Uinta Mountains. These bad lands are of gray, green, and brown shales that are carved in picturesque forms–domes, towers, pinnacles, and minarets, and bold cliffs with deep alcoves; and all are naked rock, the sediments of an ancient lake. These lake beds are filled with fossils,–the preserved bones of fishes, reptiles, and mammals, of strange and often gigantic forms, no longer found living on the globe. It is a desert to the agriculturist, a mine to the paleontologist, and a paradise to the artist.

The region thus described, from Fremont's Peak to the Uinta Mountains, has been the home of tribes of Indians of the Shoshonean family from time immemorial. It is a great hunting and fishing region, and the vigorous Shoshones still obtain a part of their livelihood from mesa and plain and river and lake. The flesh of the animals killed in fall and winter was dried in the arid winds for summer use; the trout abounding in the streams and lakes were caught at all seasons of the year; and the seeds and fruits of harvest time were gathered and preserved for winter use. When the seeds were gathered they were winnowed by tossing them in trays so that the winds might carry away the chaff. Then they were roasted in the same trays. Burning coals and seeds were mixed in the basket trays and kept in motion by a tossing process which fanned the coals until the seeds were done; then they were separated from the coals by dexterous manipulation. Afterwards the seeds were ground on mealing-stones and molded into cakes, often huge loaves, that were stored away for use in time of need. Raspberries, chokecherries, and buffalo berries are abundant, and these fruits were gathered and mixed with the bread. Such fruit cakes were great dainties among these people.

In this Shoshone land the long winter night is dedicated to worship and festival. About their camp fires scattered in forest glades by brooks and lakes, they assemble to dance and sing in honor of their gods–wonderful mythic animals, for they hold as divine the ancient of bears, the eagle of the lost centuries, the rattlesnake of primeval times, and a host of other zoic deities.

The Uinta Range stands across the course of Green River, which finds its way through it by series of stupendous canyons. The range has an east-and-west trend. The Wasatch Mountains, a long north-and-south range, here divide the Plateau Province from what is known among geologists as the Basin Range Province, on the west. The latter is the great interior basin whose waters run into salt lakes and sinks, there being no drainage to the sea. The Great Salt Lake is the most important of these interior bodies of water.

The Great Basin, which lies to the west of the Plateau Province, forms a part of the Basin Range Province. In past geological times it was the site of a vast system of lakes, but the climate has since changed and the water of most of these lakes has evaporated and the sediments of the old lake beds are now desert sands. The ancient lake shores are often represented by conspicuous terraces, each one marking a stage in the height of a dead lake. While these lakes existed the region was one of great volcanic activity and many eruptive mountains were formed. Some burst out beneath the waters; others were piled up on the dry land.

From the desert valleys below, the Wasatch Mountains rise abruptly and are crowned with craggy peaks. But on the east side of the mountains the descent to the plateau is comparatively slight. The Uinta Mountains are carved out of the great plateau which extends more than two hundred miles to the eastward of the summit of the Wasatch Range. Its mountain peaks are cameos, its upper valleys are meadows, its higher slopes are forest groves, and its streams run in deep, solemn, and majestic canyons. The snows never melt from its crowning heights, and an undying anthem is sung by its falling waters.

The Owiyukuts Plateau is situated at the northeastern end of the

Uinta Mountains. It is a great integral block of the Uinta system. A beautiful creek heads in this plateau, near its center, and descends northward into the bad lands of Vermilion Creek, to which stream it is tributary. "Once upon a time" this creek, after descending from the plateau, turned east and then southward and found its way by a beautiful canyon into Brown's Park, where it joined the Green; but a great bend of the Vermilion, near the foot of the plateau, was gradually enlarged–the stream cutting away its banks–until it encroached upon the little valley of the creek born on the Owiyukuts Plateau. This encroachment continued until at last Vermilion Creek stole the Owiyukuts Creek and carried its waters away by its own channel. Then the canyon channel through which Owiyukuts Creek had previously run, no longer having a stream to flow through its deep gorge, gathered the waters of brooks flowing along its course into little lakelets, which are connected by a running stream only through seasons of great rainfall. These lakelets in the gorge of the dead creek are now favorite resorts of Ute Indians.

South of the Uinta Mountains is the Uinta River, a stream with many mountain tributaries, some heading in the Uinta Mountains, others in the Wasatch Mountains on the west, and still others in the western Tavaputs Plateau.

The Uinta Valley is the ancient and present home of the Uinta Indians, a tribe speaking the Uinta language of the Shoshonean family. Their habits, customs, institutions, and mythology are essentially the same as those of the Ute Indians of the Grand River country, already described. In this valley there are also found many ruins of ancient pueblo-building peoples–of what stock is not known.

The Tavaputs Plateau is one of the stupendous features of this country. On the west it merges into the Wasatch Mountains; on the north it descends by wooded slopes into the Uinta Valley. Its summit is forest-clad and among the forests are many beautiful parks. On the south it ends in a great escarpment which descends into Castle Valley. This southern escarpment presents one of the most wonderful facades of the world. It is from 2,000 to 4,000 feet high. The descent is not made by one bold step, for it is cut by canyons and cliffs. It is a zone several miles in width which is a vast labyrinth of canyons, cliffs, buttes, pinnacles, minarets, and detached rocks of Cyclopean magnitude, the whole destitute of soil and vegetation, colored in many brilliant tones and tints, and carved in many weird forms,–a land of desolation, dedicated forever to the geologist and the artist, where civilization can find no resting-place.

Then comes Castle Valley, to describe which is to beggar language and pall imagination. On the north is the Tavaputs; on the west is the Wasatch Plateau, which lies to the south of the Wasatch Mountains and is here the west boundary of the Plateau Province; on the south are indescribable mesas and mountains; on the east is Grand River, a placid stream meandering through a valley of meadows. Within these boundaries there is a landscape of gigantic rock forms, interrupted here and there by bad-land hills, dominated with the towering cliffs of Tavaputs, the bold escarpment of the Wasatch Plateau, and the volcanic peaks of the Henry Mountains on the south. It is a vast forest of rock forms, and in its midst is San Rafael Swell, an elevation crowned with still more gigantic rock forms. Among the rocks pools and lakelets are found, and little streams run in canyons that seem like chasms cleft to nadir hell. San Rafael River and Fremont River drain this Castle land, heading in the Wasatch Plateau and flowing into the Grand River. Along these streams a few narrow canyon valleys are found, and in them Ute Indians make their winter homes. The bad lands are filled with agates, jaspers, and carnelians, which are gathered by the Indians and fashioned into arrowheads and knives; along the foot of the canyon cliffs workshops can be discovered that have been occupied by generations from a time in the long past, and the chips of these workshops pave the valleys. South of the Wasatch Plateau we have the Fish Lake Plateau, the Awapa Plateau, and the Aquarius Plateau, which separate the waters flowing into the Great Basin from the waters of the Colorado, which here constitute the boundary of the Plateau Province. Awapa is a Ute name signifying "Many waters."

All three of these plateaus are remarkable for the many lakelets found on them. To the east are the Henry Mountains, a group of volcanic domes that rise above the region. The rocks of the country are limestones, sandstones, and shales, originally lying in horizontal altitudes; but volcanic forces were generated under them and lavas boiled up. These lavas did not, however, come to the surface, but as they rose they lifted the sandstones, shales, and limestones, to a thickness of 2,000 or 3,000 feet or more, into great domes. Then the molten lavas cooled in great lenses of mountain magnitude, with the sedimentary rocks domed above them. Then the clouds gathered over these domes and wept, and their tears were gathered in brooks, and the brooks carved canyons down the sides of the domes; and now in these deep clefts the structure of the mountains is revealed. The lenses of volcanic rocks by which the domes were upheaved are known as "laccolites," i. e., rock lakes.

Looking southwestward from the Henry Mountains the Circle Cliffs are seen. A great escarpment, several thousand feet in height and 70 or 80 miles in length, faces the mountain. It is the step to the long, narrow plateau. The streams that come down across these cliffs head in great symmetric amphitheaters, and when first seen from above they present a vast alignment of walled circles. The front of the cliffs, seen from below, is everywhere imposing. On the southwest the Escalante River holds its course. It heads in the Aquarius Plateau and flows into the Colorado. Its course, as well as that of all its many tributaries, is in deep box-canyons of homogeneous red sandstone, often with vertical walls that are broken by many beautiful alcoves and glens. Much of the region is of naked, smooth, red rock, but the alcoves and glens that break the canyon walls are the sites of perennial springs, about which patches of luxuriant verdure gather.

The Kaiparowits Plateau is an elevated table-land on the southwestern side of the Escalante River. It is long and narrow, extending from the northwest to the southeast approximately parallel with the Escalante. It rises above the red sandstone of the Escalante region from 2,000 to 4,000 feet by a front of storm-carved cliffs. From the southeastern extremity of this plateau, at an altitude of 7,500 feet, an instructive view is obtained. One of the great canyons of the Colorado River can be seen meandering its way through the red-rock landscape. In the distance, and to the north, the Henry Mountains are in view, and below, the canyons of the Escalante and the red-rock land are in sight. Across the Colorado are the canyons of the San Juan, and below the mouth of the San Juan is the great Navajo Mountain. Still to the south the Grand Canyon of the Colorado is in view, and in the west a vast mesa landscape is presented with its buttes and pinnacles. Still to the southward Paria River is seen heading in a plateau on the margin of the province and having a course a little east of south into the Colorado.

The region of country which has been thus described, from the Tava-puts Plateau to the Paria River, was the home of a few scattered Ute Indians, who lived in very small groups, and who hunted on the plateau, fished in the waters, and dwelt in the canyons. There was nominally but one tribe, but as the members of this tribe were in very small parties and separated by wide distances the tribal bonds were very weak and often unrecognized. The chief integrating agency was religion, for they worshiped the same gods and periodically joined in the same religious ceremonies and festivals. A country so destitute of animal and vegetal life would not support large numbers, and the few who dwelt here gained but a precarious and scant subsistence. To a large extent they lived on seeds and roots. The low, warm canyons furnished admirable shelter for the people, and their habitual costumes were loincloths, paints, and necklaces of tiny arrowheads made of the bright-colored agates and carnelians strung on snakeskins.

When the Mormon people encroached on this country from the west, and when the Navajos on the east surrendered to the United States, a few recalcitrant Navajos and the Utes of this region combined. They had long been more or less intimately associated, and a jargon speech had grown up by which they could communicate. Finally, the greater number of these Utes and renegade Navajos took up their homes permanently on the eastern bank of the Colorado River between the Grand and the San Juan rivers. The Navajos are the dominant race, yet they live on terms of practical equality and affiliate without feuds. These are the great Freebooters of the Plateau Province–the enemies of other tribes and of the white men. In their canyon fortresses they have been able to hold their ground in spite of their enemies on every hand.

Throughout the region and the plateaus by which it is surrounded and the mountains by which it is interrupted, everywhere ruins of pueblos and many cliff dwellings are found. None of these ancient pueblos are on a large scale. The houses were usually one or two stories high and the hamlets rarely provided shelter for more than two dozen people. Some of the houses are of rather superior architecture, having well-constructed walls with good geometric proportions. Their houses were plastered on the inside, and sometimes on the outside, and covered with flat roofs of sun-dried mud. The real home of the people in their waking hours was on their housetops.

The rocks of the mountain are etched with many picture-writings attesting the artistic skill of this people. The predominant form is the rattlesnake, which is found in the crevices of the rocks on every hand. It is inferred that the people worshiped the rattlesnake as one of their chief deities, a god who carried the spirit of death in his mouth.

CHAPTER IV. CLIFFS AND TERRACES.

There is a great group of table-lands constituting a geographic unit which have been named the Terrace Plateaus. They ex-tend from the Paria and Colorado on the east to the Grand Wash and Pine Mountains on the west, and they are bounded on the south by the Grand Canyon of the Colorado, and on the north they divide the waters of the Colorado from the waters of the Sevier, which flows northward and then westward until it is lost in the sands of the Great Desert. It is an irregular system of great plateaus with subordinate mesas and buttes separated by lines of cliffs and dissected by canyons.

In this region all of the features which have been described as found in other portions of the province are grouped except only the cliffs of volcanic ashes, the volcanic cones, and the volcanic domes. The volcanic mountains, cinder cones, and coulees, the majestic plateaus and elaborate mesas, the sculptured buttes and canyon gorges, are all found here, but on a more stupendous scale. The volcanic mountains are higher, the cinder cones are larger, the coulees are more extensive and are often sheets of naked, black rock, the plateaus are more lofty, the cliffs are on a grander scale, the canyons are of profounder depth; and the Grand Canyon of the Colorado, the most stupendous gorge known on the globe, with a great river surging through it, bounds it on the south.

The east-and-west cliffs are escarpments of degradation, the north-and-south cliffs are, in the main, though not always, escarpments of displacement. Let us understand what this means. Over the entire region limestones, shales, and sandstones were deposited through long periods of geologic time to the thickness of many thousands of feet; then the country was upheaved and tilted toward the north; but the Colorado River was flowing when the tilting commenced, and the upheaval was very slow, so that the river cleared away the obstruction to its channel as fast as it was presented, and this is the Grand Canyon. The rocks above were carried away by rains and rivers, but not evenly all over the country; nor by washing out valleys and leaving hills, but by carving the country into terraces. The upper and later-formed rocks are found far to the north, their edges standing in cliffs; then still earlier rocks are found rising to the southward, until they terminate in cliffs; and then a third series rises to the southward and ends in cliffs, and finally a fourth series, the oldest rocks, terminating in the Grand Canyon wall, which is a line of cliffs. There are in a general way four great lines of cliffs extending from east to west across the district and presenting their faces, or escarpments, southward. If these cliffs are climbed it is found that each plateau or terrace dips gently to the northward until it meets with another line of cliffs, which must be ascended to reach the summit of another plateau. Place a book before you on a table with its front edge toward you, rest another book on the back of this, place a third on the back of the second, and in like manner a fourth on the third. Now the leaves of the books dip from you and the cut edges stand in tiny escarpments facing you. So the rock-formed leaves of these books of geology have the escarpment edges turned southward, while each book itself dips northward, and the crest of each plateau book is the summit of a line of cliffs. These cliffs of erosion have been described as running from east to west, but they diverge from that course in many ways. First, canyons run from north to south through them, and where these canyons are found deep angles occur; then sharp salients extend from the cliffs on the backs of the lower plateaus. Each great escarpment is made up more or less of minor terraces, or steps; and at the foot of each grand escarpment there is always a great talus, or sloping pile of rocks, and many marvelous buttes stand in front of the cliffs.

But these east-and-west cliffs and the plateaus which they form are divided by north-and-south lines in another manner. The country has been faulted along north-and-south lines or planes. These faults are breaks in the strata varying from 1,000 or 2,000 to 4,000 or 5,000 feet in verticality. On the very eastern margin the rocks are dropped down several thousand feet, or, which means the same thing, the rocks are upheaved on the west side; that is, the beds that were originally horizontal have been differentially displaced, so that on the west side of the fracture the strata are several thousand feet higher than they are on the east side of the fracture. The line of displacement is known as the Echo Cliff Fault.

West of this about twenty-five miles, there is another fault with its throw to the east, the upheaved rocks being on the west. This fault varies from 1,500 to 2,500 feet in throw, and extends far to the northward. It is known as the East Kaibab Fault. Still going westward, another fault is found, known as the West Kaibab Fault. Here the throw is on the west side,–that is, the rocks are dropped down to the westward from 1,000 to 2,000 feet. This fault gradually becomes less to the northward and is flexed toward the east until it joins with the East Kaibab Fault. The block between the two faults is the Kaibab Plateau. Going westward from 60 to 70 miles, still another fault is found, known as the Hurricane Ledge Fault. The throw is again on the west side of the fracture and the rocks fall down some thousands of feet. This fault extends far northward into central Utah. To the west 25 or 30 miles is found a fault with the throw still on the west. It has a drop of several thousand feet and extends across the Rio Colorado far to the southwest, probably beyond the Arizona-New Mexico line. It also extends far to the north, until it is buried and lost under the Pine Valley Mountains, which are of volcanic origin.

Now let us see what all this means. In order clearly to understand this explanation the reader is referred to the illustration designated "Section and Bird's-Eye View of the Plateaus North of the Grand Canyon." Starting at the Grand Wash on the west, the Grand Wash Cliffs, formed by the Grand Wash Fault, are scaled; and if we are but a few miles north of the Grand Canyon we are on the Shiwits Plateau. Its western boundary is the Grand Wash Cliffs, its southern boundary is the Grand Canyon, and its northern boundary is a line of cliffs of degradation, which will be described hereafter. Going eastward across the Shiwits Plateau the Hurricane Cliffs are reached, and climbing them we are on the Uinkaret Plateau, which is bounded on the south by the Grand Canyon and on the north by the Vermilion Cliffs, that rise above its northern foot. Still going eastward 30 or 40 miles to the brink of the Kanab Canyon, the West Kanab Plateau is crossed, which is bounded by the Toroweap Fault on the west, separating it from the Uinkaret Plateau, and by the Kanab Canyon on the east, with the Grand Canyon on the south and the Vermilion Cliffs on the north. Crossing the Kanab, we are on the East Kanab Plateau, which extends about 30 miles to the foot of the

West Kaibab Cliffs, or the escarpment of the West Kaibab Fault. This canyon also has the Grand Canyon on the south and the Vermilion Cliffs on the north. Climbing the West Kaibab Fault, we are on the Kaibab Plateau. Now we have been climbing from west to east, and each ascent has been made at a line of cliffs. Crossing the Kaibab Plateau to the East Kaibab Cliffs; the country falls down once more to the top of Marble Canyon Plateau. Crossing this plateau to the eastward, we at last reach the Echo Cliff Fault, where the rocks fall down on the eastern side once more; but the surface of the country itself does not fall down–the later rocks still remain, and the general level of the country is preserved except in one feature of singular interest and beauty, to describe which a little further explanation is necessary.

I have spoken of these north-and-south faults as if they were fractures; and usually they are fractures, but in some places they are flexures. The Echo Cliffs displacement is a flexure. Just over the zone of flexure a long ridge extends from north to south, known as the Echo Cliffs. It is composed of a comparatively hard and homogeneous sandstone of a later age than the limestones of the Marble Canyon Plateau west of it; but the flexure dips down so as to carry this sandstone which forms the face of the cliff (presented westward) far under the surface, so that on the east side rocks of still later age are found, the drop being several thousand feet. The inclined red sandstone stands in a ridge more than 75 miles in length, with an escarped face presented to the west and a face of inclined rock to the east. The western side is carved into beautiful alcoves and is buttressed with a magnificent talus, and the red sandstone stands in fractured columns of giant size and marvelous beauty. On the east side the declining beds are carved into pockets, which often hold water. This is the region of the Thousand Wells. The foot of the cliffs on the east side is several hundred feet above the foot of the cliffs on the west side. On the west there is a vast limestone stretch, the top of the Marble Canyon Plateau; on the east there are drifting sand-dunes.

The terraced land described has three sets of terraces: one set on the east, great steps to the Kaibab Plateau; another set on the west, from the Great Basin region to the Kaibab Plateau; and a third set from the Grand Canyon northward. There are thus three sets of cliffs: cliffs facing the east, cliffs facing the west, and cliffs facing the south. The north-and-south cliffs are made by faults; the east-and-west cliffs are made by differential degradation.

The stupendous cliffs by which the plateaus are bounded are of indescribable grandeur and beauty. The cliffs bounding the Kaibab Plateau descend on either side, and this is the culminating portion of the region. All the other plateaus are terraces, with cliffs ascending on the one side and descending on the other. Some of the tables carry dead volcanoes on their backs that are towering mountains, and all of them are dissected by canyons that are gorges of profound depth. But every one of these plateaus has characteristics peculiar to itself and is worthy of its own chapter. On the north there is a pair of plateaus, twins in age, but very distinct in development, the Paunsagunt and Markagunt. They are separated by the Sevier River, which flows northward. Their southern margins constitute the highest steps of the great system of terraces of erosion. This escarpment is known as the Pink Cliffs. Above, pine forests are found; below the cliffs are hills and sand-dunes. The cliffs themselves are bold and often vertical walls of a delicate pink color.

In one of the earlier years of exploration I stood on the summit of the Pink Cliffs of the Paunsagunt Plateau, 9,000 feet above the level of the sea. Below me, to the southwest, I could look off into the canyons of the Virgen River, down into the canyon of the Kanab, and far away into the Grand Canyon of the Colorado. From the lowlands of the Great Basin and from the depths of the Grand Canyon clouds crept up over the cliffs and floated over the landscape below me, concealing the canyons and mantling the mountains and mesas and buttes; still on toward me the clouds rolled, burying the landscape in their progress, until at last the region below was covered by a mantle of storm–a tumultuous sea of rolling clouds, black and angry in parts, white as the foam of cataracts here and there, and everywhere flecked with resplendent sheen. Below me spread a vast ocean of vapor, for I was above the clouds. On descending to the plateau, I found that a great storm had swept the land, and the dry arroyos of the day before were the channels of a thousand streams of tawny water, born of the ocean of vapor which had invaded the land before my vision.

Below the Pink Cliffs another irregular zone of plateaus is found, stretching out to the margin of the Gray Cliffs. The Gray Cliffs are composed of a homogeneous sandstone which in some places weathers gray, but in others is as white as virgin snow. On the top of these cliffs hills and sand-dunes are found, but everywhere on the Gray Cliff margin the rocks are carved in fantastic forms; not in buttes and towers and pinnacles, but in great rounded bosses of rock.

The Virgen River heads back in the Pink Cliffs of the Markagunt Plateau and with its tributaries crosses one of these plateaus above the Gray Cliffs, carving a labyrinth of deep gorges. This is known as the Colob Plateau. Above, there is a vast landscape of naked, white and gray sandstone, billowing in fantastic bosses. On the margins of the canyons these are rounded off into great vertical walls, and at the bottom of every winding canyon a beautiful stream of water is found running over quicksands. Sometimes the streams in their curving have cut under the rocks, and overhanging cliffs of towering altitudes are seen; and somber chambers are found between buttresses that uphold the walls. Among the Indians this is known as the "Rock Rovers' Land," and is peopled by mythic beings of uncanny traits.

Below the Gray Cliffs another zone of plateaus is found, separated by the north-and-south faults and divided from the Colob series by the Gray Cliffs and demarcated from the plateaus to the south by the Vermilion Cliffs. The Vermilion Cliffs that face the south are of surpassing beauty. The rocks are of orange and red above and of chocolate, lavender, gray, and brown tints below. The canyons that cut through the cliffs from north to south are of great diversity and all are of profound interest. In these canyon walls many caves are found, and often the caves contain lakelets and pools of clear water. Canyons and re-entrant angles abound. The faces of the cliffs are terraced and salients project onto the floors below. The outlying buttes are many. Standing away to the south and facing these cliffs when the sun is going down beyond the desert of the Great Basin, shadows are seen to creep into the deep recesses, while the projecting forms are illumined, so that the lights and shadows are in great and sharp contrast; then a million lights seem to glow from a background of black gloom, and a great bank of Tartarean fire stretches across the landscape.

At the foot of the Vermilion Cliffs there is everywhere a zone of vigorous junipers and pinons, for the belt of country is favored with comparatively abundant rain. When the clouds drift over the plateaus below from the south and west and strike the Vermilion Cliffs, they are abruptly lifted 2,000 feet, and to make the climb they must unload their burdens; so that here copious rains are discharged, and by such storms the cliffs are carved and ever from age to age carried back farther to the north. In the Pink Cliffs above and the Gray Cliffs and the Vermilion Cliffs, there are many notches that mark channels running northward which had their sources on these plateaus when they extended farther to the south. The Rio Virgen is the only stream heading in the Pink Cliffs and running into the Colorado which is perennial. The other rivers and creeks carry streams of water in rainy seasons only. When a succession of dry years occurs the canyons coming through the cliffs are choked below, as vast bodies of sand are deposited. But now and then, ten or twenty years apart, great storms or successions of storms come, and the channels are flooded and cut their way again through the drifting sands to solid rock below. Thus the streams below are alternately choked and cleared from period to period.

To the south of the Vermilion Cliffs the last series or zone of plateaus north of the Grand Canyon is found. The summits of these plateaus are of cherty limestone. In the far west we have the Shiwits Plateau covered with sheets of lava and volcanic cones; then climbing the Hurricane Ledge we have the Kanab Plateau, on the southwest portion of which the Uinkaret Mountains stand–a group of dead volcanoes with many black cinder cones scattered about. It is interesting to know how these mountains are formed. The first eruptions of lava were long ago, and they were poured out upon a surface 2,000 feet or more higher than the general surface now found. After the first eruptions of coulees the lands round about were degraded by rains and rivers. Then new eruptions occurred and additional sheets of lava were poured out; but these came not through the first channels, but through later ones formed about the flanks of the elder beds of lava, so that the new sheets are imbricated or shingled over the old sheets. But the overlap is from below upward. Then the land was further degraded, and a third set of coulees was spread still lower down on the flanks, and on these last coulees the black cinder cones stand. So the foundations of the Uinkaret Mountains are of limestones, and these foundations are covered with sheets of lava overlapping from below upward, and the last coulees are decked with cones.

Still farther east is the Kaibab Plateau, the culminating table-land of the region. It is covered with a beautiful forest, and in the forest charming parks are found. Its southern extremity is a portion of the wall of the Grand Canyon; its western margin is the wall of the West Kaibab Fault; its eastern edge is the wall of the East Kaibab Fault; and its northern point is found where the two faults join. Here antelope feed and many a deer goes bounding over the fallen timber. In winter deep snows lie here, but the plateau has four months of the sweetest summer man has ever known.

On the terraced plateaus three tribes of Indians are found: the Shiwits ("people of the springs"), the Uinkarets ("people of the pine mountains"), and the Unkakaniguts ("people of the red lands," who dwell along the Vermilion Cliffs). They are all Utes and belong to a confederacy with other tribes living farther to the north, in Utah. These people live in shelters made of boughs piled up in circles and covered with juniper bark supported by poles. These little houses are only large enough for half a dozen persons huddling together in sleep. Their aboriginal clothing was very scant, the most important being wildcatskin and wolfskin robes for the men, and rabbitskin robes for the women, though for occasions of festival they had clothing of tanned deer and antelope skins, often decorated with fantastic ornaments of snake skins, feathers, and the tails of squirrels and chipmunks. A great variety of seeds and roots furnish their food, and on the higher plateaus there is much game, especially deer and antelope. But the whole country abounds with rabbits, which are often killed with arrows and caught in snares. Every year they have great hunts, when scores of rabbits are killed in a single day. It is managed in this way: They make nets of the fiber of the wild flax and of some other plant, the meshes of which are about an inch across. These nets are about three and a half feet in width and hundreds of yards in length. They arrange such a net in a circle, not quite closed, supporting it by stakes and pinning the bottom firmly to the ground. From the opening of the circle they extend net wings, expanding in a broad angle several hundred yards from either side. Then the entire tribe will beat up a great district of country and drive the rabbits toward the nets, and finally into the circular snare, which is quickly closed, when the rabbits are killed with arrows.

A great variety of desert plants furnish them food, as seeds, roots, and stalks. More than fifty varieties of such seed-bearing plants have been collected. The seeds themselves are roasted, ground, and preserved in cakes. The most abundant food of this nature is derived from the sunflower and the nuts of the pinon. They still make stone arrowheads, stone knives, and stone hammers, and kindle fire with the drill. Their medicine men are famous sorcerers. Coughs are caused by invisible winged insects, rheumatism by flesh-eating bugs too small to be seen, and the toothache by invisible worms. Their healing art consists in searing and scarifying. Their medicine men take the medicine themselves to produce a state of ecstasy, in which the disease pests are discovered. They also practice dancing about their patients to drive away the evil beings or to avert the effects of sorcery. When a child is bitten by a rattlesnake the snake is caught and brought near to the suffering urchin, and ceremonies are performed, all for the purpose of prevailing upon the snake to take back the evil spirit. They have quite a variety of mythic personages. The chief of these are the Enupits, who are pigmies dwelling about the springs, and the Rock Rovers, who live in the cliffs. Their gods are zoic, and the chief among them are the wolf, the rabbit, the eagle, the jay, the rattlesnake, and the spider. They have no knowledge of the ambient air, but the winds are the breath of beasts living in the four quarters of the earth. Whirlwinds that often blow among the sand-dunes are caused by the dancing of Enupits. The sky is ice, and the rain is caused by the Rainbow God; he abraids the ice of the sky with his scales and the snow falls, and if the weather be warm the ice melts and it is rain. The sun is a poor slave compelled to make the same journey every day since he was conquered by the rabbit. These tribes have a great body of romance, in which the actors are animals, and the knowledge of these stories is the lore of their sages.

Scattered over the plateaus are the ruins of many ancient stone pueblos, not unlike those previously described.

The Kanab River heading in the Pink Cliffs runs directly southward and joins the Colorado in the heart of the Grand Canyon. Its way is through a series of canyons. From one of these it emerges at the foot of the Vermilion Cliffs, and here stood an extensive ruin not many years ago. Some portions of the pueblo were three stories high. The structure was one of the best found in this land of ruins. The Mormon people settling here have used the stones of the old pueblo in building their homes, and now no vestiges of the ancient structure remain. A few miles below the town other ruins were found. They were scattered to Pipe's Springs, a point twenty miles to the westward. Ruins were also discovered up the stream as far as the Pink Cliffs, and eastward along the Vermilion Cliffs nearly to the Colorado River, and out on the margin of the Kanab Plateau. These were all ruins of outlying habitations be-longing to the Kanab pueblo. From the study of the existing pueblos found elsewhere and from extensive study of the ruins, it seems that everywhere tribal pueblos were built of considerable dimensions, usually to give shelter to several hundred people. Then the people cultivated the soil by irrigation, and had their gardens and little fields scattered at wide distances about the central pueblo, by little springs and streams and wherever they could control the water with little labor to bring it on the land. At such points stone houses were erected sufficient to accommodate from one to two thousand people, and these were occupied during the season of cultivation and are known as rancherias. So one great tribe had its central pueblo and its outlying rancherias. Sometimes the rancherias were occupied from year to year, especially in time of peace, but usually they were occupied only during seasons of cultivation. Such groups of ruins and pueblos with accessory rancherias are still inhabited, and have been described as found throughout the Plateau Province except far to the north beyond the Uinta Mountains. A great pueblo once existed in the Uinta Valley on the south side of the mountains. This is the most northern pueblo which has yet been discovered. But the pueblo-building tribes extended beyond the area drained by the Colorado. On the west there was a pueblo in the Great Basin at the site now occupied by Salt Lake City, and several more to the southward, all on waters flowing into the desert. On the east such pueblos were found among mountains at the headwaters of the Arkansas, Platte, and Canadian rivers. The entire area drained by the Rio Grande del Norte was occupied by pueblo tribes, and a number are still inhabited. To the south they extended far beyond the territory of the United States, and the so-called Aztec cities were rather superior pueblos of this character. The known pueblo tribes of the United States belong to several different linguistic stocks. They are far from being one homogeneous people, for they have not only different languages but different religions and worship different gods. These pueblo peoples are in a higher grade of culture than most Indian tribes of the United States. This is exhibited in the slight superiority of their arts, especially in their architecture. It is also noticeable in their mythology and religion. Their gods, the heroes of their myths, are more often personifications of the powers and phenomena of nature, and their religious ceremonies are more elaborate, and their cult societies are highly organized. As they had begun to domesticate animals and to cultivate the soil, so as to obtain a part of their subsistence by agriculture, they had almost accomplished the ascent from savagery to barbarism when first discovered by the invading European. All the Indians of North America were in this state of transition, but the pueblo tribes had more nearly reached the higher goal.

The great number of ruins found throughout the land has often been interpreted as evidence of a much larger pueblo population than has been found in post-Columbian time. But a careful study of the facts does not warrant this conclusion. It would seem that for various reasons tribes abandoned old pueblos and built new, thus changing their permanent residence from time to time; but more frequent changes were made in their rancherias. These were but ephemeral, being moved from place to place by the varying conditions of water supply. Most of the streams of the arid land are not perennial, but very many of the smaller streams of the pueblo region discharge their waters into the larger streams in times of great flood. Such floods occur now here, now there, and at varying periods, sometimes fifty years apart. When dry years follow one another for a long series, the channels of these intermittent streams are choked with sand until the streams are buried and lost. Under such circumstances the rancherias were moved from dead stream to living stream. In rare instances pueblos themselves were removed for this cause. Other pueblos, and the rancherias generally, were abandoned in time of war; this seems to have been a potent cause for moving. When pestilence attacked a pueblo the people would sometimes leave in a body and never return. The cliff pueblos and dwellings, the cavate dwellings, and the cinder-cone towns were all built and occupied for defensive purposes when powerful enemies threatened. The history of some of the old ruins has been obtained and we know the existing tribes who once occupied them; others still remain enshrouded in obscurity.

CHAPTER V. FROM GREEN RIVER CITY TO FLAMING GORGE.

In the summer of 1867, with a small party of naturalists, students, and amateurs like myself, I visited the mountain region of Colorado Territory. While in Middle Park I explored a little canyon through which the Grand River runs, immediately below the now well-known watering place, Middle Park Hot Springs. Later in the fall I passed through Cedar Canyon, the gorge by which the Grand leaves the park. A result of the summer's study was to kindle a desire to explore the canyons of the Grand, Green, and Colorado rivers, and the next summer I organized an expedition with the intention of penetrating still farther into that canyon country.

As soon as the snows were melted, so that the main range could be crossed, I went over into Middle Park, and proceeded thence down the Grand to the head of Cedar Canyon, then across the Park Range by Gore's Pass, and in October found myself and party encamped on the White River, about 120 miles above its mouth. At that point I built cabins and established winter quarters, intending to occupy the cold season, as far as possible, in exploring the adjacent country. The winter of 1868-69 proved favorable to my purposes, and several excursions were made, southward to the Grand, down the White to the Green, northward to the Yampa, and around the Uinta Mountains. During these several excursions I seized every opportunity to study the canyons through which these upper streams run, and while thus engaged formed plans for the exploration of the canyons of the Colorado. Since that time I have been engaged in executing these plans, sometimes employed in the field, sometimes in the office. Begun originally as an exploration, the work was finally developed into a survey, embracing the geography, geology, ethnography, and natural history of the country, and a number of gentlemen have, from time to time, assisted me in the work.

Early in the spring of 1869 a party was organized for the exploration of the canyons. Boats were built in Chicago and transported by rail to the point where the Union Pacific Railroad crosses the Green River. With these we were to descend the Green to the Colorado, and the Colorado down to the foot of the Grand Canyon.

May 24, 1869.–The good people of Green River City turn out to see us start. We raise our little flag, push the boats from shore, and the swift current carries us down.

Our boats are four in number. Three are built of oak; stanch and firm; double-ribbed, with double stem and stern posts, and further strengthened by bulkheads, dividing each into three compartments. Two of these, the fore and aft, are decked, forming water-tight cabins. It is expected these will buoy the boats should the waves roll over them in rough water. The fourth boat is made of pine, very light, but 16 feet in length, with a sharp cutwater, and every way built for fast rowing, and divided into compartments as the others. The little vessels are 21 feet long, and, taking out the cargoes, can be carried by four men.

We take with us rations deemed sufficient to last ten months, for we expect, when winter comes on and the river is filled with ice, to lie over at some point until spring arrives; and so we take with us abundant supplies of clothing, likewise. We have also a large quantity of ammunition and two or three dozen traps. For the purpose of building cabins, repairing boats, and meeting other exigencies, we are supplied with axes, hammers, saws, augers, and other tools, and a quantity of nails and screws. For scientific work, we have two sextants, four chronometers, a number of barometers, thermometers, compasses, and other instruments.

The flour is divided into three equal parts; the meat, and all other articles of our rations, in the same way. Each of the larger boats has an axe, hammer, saw, auger, and other tools, so that all are loaded alike. We distribute the cargoes in this way that we may not be entirely destitute of some important article should any one of the boats be lost. In the small boat we pack a part of the scientific instruments, three guns, and three small bundles of clothing, only; and in this I proceed in advance to explore the channel.

J. C. Sumner and William H. Dunn are my boatmen in the "Emma Dean"; then follows "Kitty Clyde's Sister," manned by W. H. Powell and G. Y. Bradley; next, the "No Name," with O. G. Howland, Seneca Howland, and Frank Goodman; and last comes the "Maid of the Canyon," with W. E. Hawkins and Andrew Hall.

Sumner was a soldier during the late war, and before and since that time has been a great traveler in the wilds of the Mississippi Valley and the Rocky Mountains as an amateur hunter. He is a fair-haired, delicate-looking man, but a veteran in experience, and has performed the feat of crossing the Rocky Mountains in midwinter on snowshoes. He spent the winter of 1886-87 in Middle Park, Colorado, for the purpose of making some natural history collections for me, and succeeded in killing three grizzlies, two mountain lions, and a large number of elk, deer, sheep, wolves, beavers, and many other animals. When Bayard Taylor traveled through the parks of Colorado, Sumner was his guide, and he speaks in glowing terms of Mr. Taylor's genial qualities in camp, but he was mortally offended when the great traveler requested him to act as doorkeeper at Breckenridge to receive the admission fee from those who attended his lectures.

Dunn was a hunter, trapper, and mule-packer in Colorado for many years. He dresses in buckskin with a dark oleaginous luster, doubtless due to the fact that he has lived on fat venison and killed many beavers since he first donned his uniform years ago. His raven hair falls down to his back, for he has a sublime contempt of shears and razors.

Captain Powell was an officer of artillery during the late war and was captured on the 22d day of July, 1864, at Atlanta and served a ten months' term in prison at Charleston, where he was placed with other officers under fire. He is silent, moody, and sarcastic, though sometimes he enlivens the camp at night with a song. He is never surprised at anything, his coolness never deserts him, and he would choke the belching throat of a volcano if he thought the spitfire meant anything but fun. We call him "Old Shady."

Bradley, a lieutenant during the late war, and since orderly sergeant in the regular army, was, a few weeks previous to our start, discharged, by order of the Secretary of War, that he might go on this trip. He is scrupulously careful, and a little mishap works him into a passion, but when labor is needed he has a ready hand and powerful arm, and in danger, rapid judgment and unerring skill. A great difficulty or peril changes the petulant spirit into a brave, generous soul.

O. G. Howland is a printer by trade, an editor by profession, and a hunter by choice. When busily employed he usually puts his hat in his pocket, and his thin hair and long beard stream in the wind, giving him a wild look, much like that of King Lear in an illustrated copy of Shakespeare which tumbles around the camp.

Seneca Howland is a quiet, pensive young man, and a great favorite with all.

Goodman is a stranger to us–a stout, willing Englishman, with florid face and more florid anticipations of a glorious trip.

Billy Hawkins, the cook, was a soldier in the Union Army during the war, and when discharged at its close went West, and since then has been engaged as teamster on the plains or hunter in the mountains. He is an athlete and a jovial good fellow, who hardly seems to know his own strength.

Hall is a Scotch boy, nineteen years old, with what seems to us a "secondhand head," which doubtless came down to him from some knight who wore it during the Border Wars. It looks a very old head indeed, with deep-set blue eyes and beaked nose. Young as he is, Hall has had experience in hunting, trapping, and fighting Indians, and he makes the most of it, for he can tell a good story, and is never encumbered by unnecessary scruples in giving to his narratives those embellishments which help to make a story complete. He is always ready for work or play and is a good hand at either.

Our boats are heavily loaded, and only with the utmost care is it possible to float in the rough river without shipping water. A mile or two below town we run on a sandbar. The men jump into the stream and thus lighten the vessels, so that they drift over, and on we go.

In trying to avoid a rock an oar is broken on one of the boats, and, thus crippled, she strikes. The current is swift and she is sent reeling and rocking into the eddy. In the confusion two other oars are lost overboard, and the men seem quite discomfited, much to the amusement of the other members of the party. Catching the oars and starting again, the boats are once more borne down the stream, until we land at a small cottonwood grove on the bank and camp for noon.

During the afternoon we run down to a point where the river sweeps the foot of an overhanging cliff, and here we camp for the night. The sun is yet two hours high, so I climb the cliffs and walk back among the strangely carved rocks of the Green River bad lands. These are sandstones and shales, gray and buff, red and brown, blue and black strata in many alternations, lying nearly horizontal, and almost without soil and vegetation. They are very friable, and the rain and streams have carved them into quaint shapes. Barren desolation is stretched before me; and yet there is a beauty in the scene. The fantastic carvings, imitating architectural forms and suggesting rude but weird statuary, with the bright and varied colors of the rocks, conspire to make a scene such as the dweller in verdure-clad hills can scarcely appreciate.

Standing on a high point, I can look off in every direction over a vast landscape, with salient rocks and cliffs glittering in the evening sun. Dark shadows are settling in the valleys and gulches, and the heights are made higher and the depths deeper by the glamour and witchery of light and shade. Away to the south the Uinta Mountains stretch in a long line,–high peaks thrust into the sky, and snow fields glittering like lakes of molten silver, and pine forests in somber green, and rosy clouds playing around the borders of huge, black masses; and heights and clouds and mountains and snow fields and forests and rock-lands are blended into one grand view. Now the sun goes down, and I return to camp.

May 25.–We start early this morning and run along at a good rate until about nine o'clock, when we are brought up on a gravelly bar. All jump out and help the boats over by main strength. Then a rain comes on, and river and clouds conspire to give us a thorough drenching. Wet, chilled, and tired to exhaustion, we stop at a cottonwood grove on the bank, build a huge fire, make a cup of coffee, and are soon refreshed and quite merry. When the clouds "get out of our sunshine" we start again. A few miles farther down a flock of mountain sheep are seen on a cliff to the right. The boats are quietly tied up and three or four men go after them. In the course of two or three hours they return. The cook has been successful in bringing down a fat lamb. The unsuccessful hunters taunt him with finding it dead; but it is soon dressed, cooked, and eaten, and makes a fine four o'clock dinner.

"All aboard," and down the river for another dozen miles. On the way we pass the mouth of Black's Fork, a dirty little stream that seems somewhat swollen. Just below its mouth we land and camp.

May 26.–To-day we pass several curiously shaped buttes, standing between the west bank of the river and the high bluffs beyond. These buttes are outliers of the same beds of rocks as are exposed on the faces of the bluffs,–thinly laminated shales and sandstones of many colors, standing above in vertical cliffs and buttressed below with a water-carved talus; some of them attain an altitude of nearly a thousand feet above the level of the river.

We glide quietly down the placid stream past the carved cliffs of the mauvaises terres, now and then obtaining glimpses of distant mountains. Occasionally, deer are started from the glades among the willows; and several wild geese, after a chase through the water, are shot. After dinner we pass through a short and narrow canyon into a broad valley; from this, long, lateral valleys stretch back on either side as far as the eye can reach.

Two or three miles below, Henry's Fork enters from the right. We land a short distance above the junction, where a cache of instruments and rations was made several months ago in a cave at the foot of the cliff, a distance back from the river. Here they were safe from the elements and wild beasts, but not from man. Some anxiety is felt, as we have learned that a party of Indians have been camped near the place for several weeks. Our fears are soon allayed, for we find the cache undisturbed. Our chronometer wheels have not been taken for hair ornaments, our barometer tubes for beads, or the sextant thrown into the river as "bad medicine," as had been predicted. Taking up our cache, we pass down to the foot of the Uinta Mountains and in a cold storm go into camp.

The river is running to the south; the mountains have an easterly and westerly trend directly athwart its course, yet it glides on in a quiet way as if it thought a mountain range no formidable obstruction. It enters the range by a flaring, brilliant red gorge, that may be seen from the north a score of miles away. The great mass of the mountain ridge through which the gorge is cut is composed of bright vermilion rocks; but they are surmounted by broad bands of mottled buff and gray, and these bands come down with a gentle curve to the water's edge on the nearer slope of the mountain.

This is the head of the first of the canyons we are about to explore–an introductory one to a series made by the river through this range. We name it Flaming Gorge. The cliffs, or walls, we find on measurement to be about 1,200 feet high.

May 27.–To-day it rains, and we employ the time in repairing one of our barometers, which was broken on the way from New York. A new tube has to be put in; that is, a long glass tube has to be filled with mercury, four or five inches at a time, and each installment boiled over a spirit lamp. It is a delicate task to do this without breaking the glass; but we have success, and are ready to measure mountains once more.

May 28.–To-day we go to the summit of the cliff on the left and take observations for altitude, and are variously employed in topographic and geologic work.

May 29.–This morning Bradley and I cross the river and climb more than a thousand feet to a point where we can see the stream sweeping in a long, beautiful curve through the gorge below.

Turning and looking to the west, we can see the valley of Henry's Fork, through which, for many miles, the little river flows in a tortuous channel. Cottonwood groves are planted here and there along its course, and between them are stretches of grass land. The narrow mountain valley is inclosed on either side by sloping walls of naked rock of many bright colors. To the south of the valley are the Uintas, and the peaks of the Wasatch Mountains can be faintly seen in the far west. To the north, desert plains, dotted here and there with curiously carved hills and buttes, extend to the limit of vision.

For many years this valley has been the home of a number of mountaineers, who were originally hunters and trappers, living with the Indians. Most of them have one or more Indian wives. They no longer roam with the nomadic tribes in pursuit of buckskin or beaver, but have accumulated herds of cattle and horses, and consider themselves quite well to do. Some of them have built cabins; others still live in lodges. John Baker is one of the most famous of these men, and from our point of view we can see his lodge, three or four miles up the river.

The distance from Green River City to Flaming Gorge is 62 miles. The river runs between bluffs, in some places standing so close to each other that no flood plain is seen. At such a point the river might properly be said to run through a canyon. The bad lands on either side are interrupted here and there by patches of Artemisia, or sage brush. Where there is a flood plain along either side of the river, a few cottonwoods may be seen.

CHAPTER VI. FROM FLAMING GORGE TO THE GATE OF LODORE.

One must not think of a mountain range as a line of peaks standing on a plain, but as a broad platform many miles wide from which mountains have been carved by the waters. One must conceive, too, that this plateau is cut by gulches and canyons in many directions and that beautiful valleys are scattered about at different altitudes. The first series of canyons we are about to explore constitutes a river channel through such a range of mountains. The canyon is cut nearly halfway through the range, then turns to the east and is cut along the central line, or axis, gradually crossing it to the south. Keeping this direction for more than 50 miles, it then turns abruptly to a southwest course, and goes diagonally through the southern slope of the range.

This much we know before entering, as we made a partial exploration of the region last fall, climbing many of its peaks, and in a few places reaching the brink of the canyon walls and looking over precipices many hundreds of feet high to the water below.

Here and there the walls are broken by lateral canyons, the channels of little streams entering the river. Through two or three of these we found our way down to the Green in early winter and walked along the low water-beach at the foot of the cliffs for several miles. Where the river has this general easterly direction the western part only has cut for itself a canyon, while the eastern has formed a broad valley, called, in honor of an old-time trapper, Brown's Park, and long known as a favorite winter resort for mountain men and Indians.

May 30.–This morning we are ready to enter the mysterious canyon, and start with some anxiety. The old mountaineers tell us that it cannot be run; the Indians say, "Water heap catch 'em"; but all are eager for the trial, and off we go.

Entering Flaming Gorge, we quickly run through it on a swift current and emerge into a little park. Half a mile below, the river wheels sharply to the left and enters another canyon cut into the mountain. We enter the narrow passage. On either side the walls rapidly increase in altitude. On the left are overhanging ledges and cliffs,–500, 1,000, 1,500 feet high.

On the right the rocks are broken and ragged, and the water fills the channel from cliff to cliff. Now the river turns abruptly around a point to the right, and the waters plunge swiftly down among great rocks; and here we have our first experience with canyon rapids. I stand up on the deck of my boat to seek a way among the wave-beaten rocks. All untried as we are with such waters, the moments are filled with intense anxiety. Soon our boats reach the swift current; a stroke or two, now on this. side, now on that, and we thread the narrow passage with exhilarating Velocity, mounting the high waves, whose foaming crests dash over us, and plunging into the troughs, until we reach the quiet water below. Then comes a feeling of great relief. Our first rapid is run. Another mile, and we come into the valley again.

Let me explain this canyon. Where the river turns to the left above, it takes a course directly into the mountain, penetrating to its very heart, then wheels back upon itself, and runs out into the valley from which it started only half a mile below the point at which it entered; so the canyon is in the form of an elongated letter U, with the apex in the center of the mountain. We name it Horseshoe Canyon.

Soon we leave the valley and enter another short canyon, very narrow at first, but widening below as the canyon walls increase in height. Here we discover the mouth of a beautiful little creek coming down through its narrow water-worn cleft. Just at its entrance there is a park of two or three hundred acres, walled on every side by almost vertical cliffs hundreds of feet in altitude, with three gateways through the walls–one up the river, another down, and a third through which the creek comes in. The river is broad, deep, and quiet, and its waters mirror towering rocks.

Kingfishers are playing about the streams, and so we adopt as names Kingfisher Creek, Kingfisher Park, and Kingfisher Canyon.

At night we camp at the foot of this canyon.

Our general course this day has been south, but here the river turns to the east around a point which is rounded to the shape of a dome. On its sides little cells have been carved by the action of the water, and in these pits, which cover the face of the dome, hundreds of swallows have built their nests. As they flit about the cliffs, they look like swarms of bees, giving to the whole the appearance of a colossal beehive of the old-time form, and so we name it Beehive Point.

The opposite wall is a vast amphitheater, rising in a succession of terraces to a height of 1,200 or 1,500 feet. Each step is built of red sandstone, with a face of naked red rock and a glacis clothed with verdure. So the amphitheater seems banded red and green, and the evening sun is playing with roseate flashes on the rocks, with shimmering green on the cedars' spray, and with iridescent gleams on the dancing waves. The landscape revels in the sunshine.

May 31.–We start down another canyon and reach rapids made dangerous by high rocks lying in the channel; so we run ashore and let our boats down with lines. In the afternoon we come to more dangerous rapids and stop to examine them. I find we must do the same work again, but, being on the wrong side of the river to obtain a foothold, must first cross over–no very easy matter in such a current, with rapids and rocks below. We take the pioneer boat, "Emma Dean," over, and unload her on the bank; then she returns and takes another load. Running back and forth, she soon has half our cargo over. Then one of the larger boats is manned and taken across, but is carried down almost to the rocks in spite of hard rowing. The other boats follow and make the landing, and we go into camp for the night.

At the foot of the cliff on this side there is a long slope covered with pines; under these we make our beds, and soon after sunset are seeking rest and sleep. The cliffs on either side are of red sandstone and stretch toward the heavens 2,500 feet. On this side the long, pine-clad slope is surmounted by perpendicular cliffs, with pines on their summits. The wall on the other side is bare rock from the water's edge up 2,000 feet, then slopes back, giving footing to pines and cedars.

As the twilight deepens, the rocks grow dark and somber; the threatening roar of the water is loud and constant, and I lie awake with thoughts of the morrow and the canyons to come, interrupted now and then by characteristics of the scenery that attract my attention. And here I make a discovery. On looking at the mountain directly in front, the steepness of the slope is greatly exaggerated, while the distance to its summit and its true altitude are correspondingly diminished. I have heretofore found that to judge properly of the slope of a mountain side, one must see it in profile. In coming down the river this afternoon, I observed the slope of a particular part of the wall and made an estimate of its altitude. While at supper, I noticed the same cliff from a position facing it, and it seemed steeper, but not half so high. Now lying on my side and looking at it, the true proportions appear. This seems a wonder, and I rise to take a view of it standing. It is the same cliff as at supper time. Lying down again, it is the cliff as seen in profile, with a long slope and distant summit. Musing on this, I forget "the morrow and the canyons to come"; I have found a way to estimate the altitude and slope of an inclination, in like manner as I can judge of distance along the horizon. The reason is simple. A reference to the stereoscope will suggest it. The distance between the eyes forms a base line for optical triangulation.

June 1.–To-day we have an exciting ride. The river rolls down the canyon at a wonderful rate, and, with no rocks in the way, we make almost railroad speed. Here and there the water rushes into a narrow gorge; the rocks on the side roll it into the center in great waves, and the boats go leaping and bounding over these like things of life, reminding me of scenes witnessed in Middle Park–herds of startled deer bounding through forests beset with fallen timber. I mention the resemblance to some of the hunters, and so striking is it that the expression, "See the blacktails jumping the logs," comes to be a common one. At times the waves break and roll over the boats, which necessitates much bailing and obliges us to stop occasionally for that purpose. At one time we run twelve miles in an hour, stoppages included.

Last spring I had a conversation with an old Indian named Pariate, who told me about one of his tribe attempting to run this canyon. "The rocks," he said, holding his hands above his head, his arms vertical, and looking between them to the heavens, "the rocks h-e-a-p,

OVEN NEAR PESCADO PUEBLO.

h-e-a-p high; the water go h-oo-woogh, h-oo-woogh; water-pony li-e-a-p buck; water catch 'em; no see 'em Injun any more! no see 'em squaw any more! no see 'em papoose any more!"

Those who have seen these wild Indian ponies rearing alternately before and behind, or "bucking," as it is called in the vernacular, will appreciate his description.

At last we come to calm water, and a threatening roar is heard in the distance. Slowly approaching the point whence the sound issues, we come near to falls, and tie up just above them on the left. Here we shall be compelled to make a portage; so we unload the boats, and fasten a long line to the bow of the smaller one, and another to the stern, and moor her close to the brink of the fall. Then the bowline is taken below and made fast; the stern line is held by five or six men, and the boat let down as long as they can hold her against the rushing waters; then, letting go one end of the line, it runs through the ring; the boat leaps over the fall and is caught by the lower rope.

Now we rest for the night.

June 2.–This morning we make a trail among the rocks, transport the cargoes to a point below the fall, let the remaining boats over, and are ready to start before noon.

On a high rock by which the trail passes we find the inscription: "Ashley 18-5." The third figure is obscure–some of the party reading it 1835, some 1855. James Baker, an old-time mountaineer, once told me about a party of men starting down the river, and Ashley was named as one. The story runs that the boat was swamped, and some of the party drowned in one of the canyons below. The word "Ashley" is a warning to us, and we resolve on great caution. Ashley Falls is the name we give to the cataract.

The river is very narrow, the right wall vertical for 200 or 300 feet, the left towering to a great height, with a vast pile of broken rocks lying between the foot of the cliff and the water. Some of the rocks broken down from the ledge above have tumbled into the channel and caused this fall. One great cubical block, thirty or forty feet high, stands in the middle of the stream, and the waters, parting to either side, plunge down about twelve feet, and are broken again by the smaller rocks into a rapid below. Immediately below the falls the water occupies the entire channel, there being no talus at the foot of the cliffs.

We embark and run down a short distance, where we find a landing-place for dinner.

On the waves again all the afternoon. Near the lower end of this canyon, to which we have given the name of Red Canyon, is a little park, where streams come down from distant mountain summits and enter the river on either side; and here we camp for the night under two stately pines.

June 3.–This morning we spread our rations, clothes, etc., on the ground to dry, and several of the party go out for a hunt. I take a walk of five or six miles up to a pine-grove park, its grassy carpet bedecked with crimson velvet flowers, set in groups on the stems of pear-shaped cactus plants; patches of painted cups are seen here and there, with yellow blossoms protruding through scarlet bracts; little blue-eyed flowers are peeping through the grass; and the air is filled with fragrance from the white blossoms of the Spiraea. A mountain brook runs through the midst, ponded below by beaver dams. It is a quiet place for retirement from the raging waters of the canyon.

It will be remembered that the course of the river from Flaming Gorge to Beehive Point is in a southerly direction and at right angles to the Uinta Mountains, and cuts into the range until it reaches a point within five miles of the crest, where it turns to the east and pursues a course not quite parallel to the trend of the range, but crosses the axis slowly in a direction a little south of east. Thus there is a triangular tract between the river and the axis of the mountain, with its acute angle extending eastward. I climb the mountain overlooking this country. To the east the peaks are not very high, and already most of the snow has melted, but little patches lie here and there under the lee of ledges of rock. To the west the peaks grow higher and the snow fields larger. Between the brink of the canyon and the foot of these peaks, there is a high bench. A number of creeks have their sources in the snowbanks to the south and run north into the canyon, tumbling down from 3,000 to 5,000 feet in a distance of five or six miles. Along their upper courses they run through grassy valleys, but as they approach Red Canyon they rapidly disappear under the general surface of the country, and emerge into the canyon below in deep, dark gorges of their own. Each of these short lateral canyons is marked by a succession of cascades and a wild confusion of rocks and trees and fallen timber and thick undergrowth.

The little valleys above are beautiful parks; between the parks are stately pine forests, half hiding ledges of red sandstone. Mule deer and elk abound; grizzly bears, too, are abundant; and here wild cats, wolverines, and mountain lions are at home. The forest aisles are filled with the music of birds, and the parks are decked with flowers. Noisy brooks meander through them; ledges of moss-covered rocks are seen; and gleaming in the distance are the snow fields, and the mountain tops are away in the clouds.

June 4--We start early and run through to Brown's Park. Halfway down the valley, a spur of a red mountain stretches across the river, which cuts a canyon through it. Here the walls are comparatively low, but vertical. A vast number of swallows have built their adobe houses on the face of the cliffs, on either side of the river. The waters are deep and quiet, but the swallows are swift and noisy enough, sweeping by in their curved paths through the air or chattering from the rocks, while the young ones stretch their little heads on naked necks through the doorways of their mud houses and clamor for food. They are a noisy people. We call this Swallow Canyon.

Still down the river we glide until an early hour in the afternoon, when we go into camp under a giant cottonwood standing on the right bank a little way back from the stream. The party has succeeded in killing a fine lot of wild ducks, and during the afternoon a mess of fish is taken.

June 5.–With one of the men I climb a mountain, off on the right. A long spur, with broken ledges of rock, puts down to the river, and along its course, or up the "hogback," as it is called, I make the ascent. Dunn, who is climbing to the same point, is coming up the gulch. Two hours' hard work has brought us to the summit. These mountains are all verdure-clad; pine and cedar forests are set on green terraces; snow-clad mountains are seen in the distance, to the west; the plains of the upper Green stretch out before us to the north until they are lost in the blue heavens; but half of the river-cleft range intervenes, and the river itself is at our feet.

This half range, beyond the river, is composed of long ridges nearly parallel with the valley. On the farther ridge, to the north, four creeks have their sources. These cut through the intervening ridges, one of which is much higher than that on which they head, by canyon gorges; then they run with gentle curves across the valley, their banks set with willows, box-elders, and cottonwood groves. To the east we look up the valley of the Vermilion, through which Fremont found his path on his way to the great parks of Colorado.

The reading of the barometer taken, we start down in company, and reach camp tired and hungry, which does not abate one bit our enthusiasm as we tell of the day's work with its glory of landscape.

June 6.–At daybreak I am awakened by a chorus of birds. It seems as if all the feathered songsters of the region have come to the old tree. Several species of warblers, woodpeckers, and flickers above, meadow larks in the grass, and wild geese in the river. I recline on my elbow and watch a lark near by, and then awaken my bedfellow, to listen to my Jenny Lind. A real morning concert for me; none of your "matinees"!

Our cook has been an ox-driver, or "bull-whacker," on the plains, in one of those long trains now no longer seen, and he hasn't forgotten his old ways. In the midst of the concert, his voice breaks in: "Roll out! roll out! bulls in the corral! chain up the gaps! Roll out! roll out! roll out!" And this is our breakfast bell.

To-day we pass through, the park, and camp at the head of another canyon.

June 7.–To-day two or three of us climb to the summit of the cliff on the left, and find its altitude above camp to be 2,086 feet. The rocks are split with fissures, deep and narrow, sometimes a hundred feet or more to the bottom, and these fissures are filled with loose earth and decayed vegetation in which lofty pines find root. On a rock we find a pool of clear, cold water, caught from yesterday evening's shower. After a good drink we walk out to the brink of the canyon and look down to the water below. I can do this now, but it has taken several years of mountain climbing to cool my nerves so that I can sit with my feet over the edge and calmly look down a precipice 2,000 feet. And yet I cannot look on and see another do the same. I must either bid him come away or turn my head. The canyon walls are buttressed on a grand scale, with deep alcoves intervening; columned crags crown the cliffs, and the river is rolling below.

When we return to camp at noon the sun shines in splendor on vermilion walls, shaded into green and gray where the rocks are lichened over; the river fills the channel from wall to wall, and the canyon opens, like a beautiful portal, to a region of glory. This evening, as I write, the sun is going down and the shadows are settling in the canyon. The vermilion gleams and roseate hues, blending with the green and gray tints, are slowly changing to somber brown above, and black shadows are creeping over them below; and now it is a dark portal to a region of gloom–the gateway through which we are to enter on our voyage of exploration tomorrow. What shall we find?

The distance from Flaming Gorge to Beehive Point is 9 2/3 miles. Besides passing through the gorge, the river runs through Horseshoe and Kingfisher canyons, separated by short valleys. The highest point on the walls at Flaming Gorge is 1,300 feet above the river. The east wall at the apex of Horseshoe Canyon is about 1,600 feet above the water's edge, and from this point the walls slope both to the head and foot of the canyon.

Kingfisher Canyon, starting at the water's edge above, steadily increases in altitude to 1,200 feet at the foot.

Red Canyon is 25 2/3 miles long, and the highest walls are about 2,500 feet.

Brown's Park is a valley, bounded on either side by a mountain range, really an expansion of the canyon. The river, through the park, is 35 1/2 miles long, but passes through two short canyons on its way, where spurs from the mountains on the south are thrust across its course.

CHAPTER VII. THE CANYON OF LODORE.

June 8.–We enter the canyon, and until noon find a succession of rapids, over which, our boats have to be taken. Here I must explain our method of proceeding at such places. The "Emma Dean "'goes in advance; the other boats follow, in obedience to signals. When we approach a rapid, or what on other rivers would often be called a fall, I stand on deck to examine it, while the oarsmen back water, and we drift on as slowly as possible. If I can see a clear chute between the rocks, away we go; but if the channel is beset entirely across, we signal the other boats, pull to land, and I walk along the shore for closer examination. If this reveals no clear channel, hard work begins. We drop the boats to the very head of the dangerous place and let them over by lines or make a portage, frequently carrying both boats and cargoes over the rocks.

The waves caused by such falls in a river differ much from the waves of the sea. The water of an ocean wave merely rises and falls; the form only passes on, and form chases form unceasingly. A body floating on such waves merely rises and sinks–does not progress unless impelled by wind or some other power. But here the water of the wave passes on while the form remains. The waters plunge down ten or twenty feet to the foot of a fall, spring up again in a great wave, then down and up in a series of billows that gradually disappear in the more quiet waters below; but these waves are always there, and one can stand above and count them.

A boat riding such billows leaps and plunges along with great velocity. Now, the difficulty in riding over these falls, when no rocks are in the way, is with the first wave at the foot. This will sometimes gather for a moment, heap up higher and higher, and then break back.

If the boat strikes it the instant after it breaks, she cuts through, and the mad breaker dashes its spray over the boat and washes overboard all who do not cling tightly. If the boat, in going over the falls, chances to get caught in some side current and is turned from its course, so as to strike the wave "broadside on," and the wave breaks at the same instant, the boat is capsized; then we must cling to her, for the water-tight compartments act as buoys and she cannot sink; and so we go, dragged through the waves, until still waters are reached, when we right the boat and climb aboard. We have several such experiences to-day.

At night we camp on the right bank, on a little shelving rock between the river and the foot of the cliff; and with night comes gloom into these great depths. After supper we sit by our camp fire, made of driftwood caught by the rocks, and tell stories of wild life; for the men have seen such in the mountains or on the plains, and on the battlefields of the South. It is late before we spread our blankets on the beach.

Lying down, we look up through the canyon and see that only a little of the blue heaven appears overhead–a crescent of blue sky, with two or three constellations peering down upon us. I do not sleep for some time, as the excitement of the day has not worn off. Soon I see a bright star that appears to rest on the very verge of the cliff overhead to the east. Slowly it seems to float from its resting place on the rock over the canyon. At first it appears like a jewel set on the brink of the cliff, but as it moves out from the rock I almost wonder that it does not fall. In fact, it does seem to descend in a gentle curve, as though the bright sky in which the stars are set were spread across the canyon, resting on either wall, and swayed down by its own weight. The stars appear to be in the canyon. I soon discover that it is the bright star Vega; so it occurs to me to designate this part of the wall as the "Cliff of the Harp."

June 9.–One of the party suggests that we call this the Canyon of Lodore, and the name is adopted. Very slowly we make our way, often climbing on the rocks at the edge of the water for a few hundred yards to examine the channel before running it. During the afternoon we come to a place where it is necessary to make a portage. The little boat is landed and the others are signaled to come up.

When these rapids or broken falls occur usually the channel is suddenly narrowed by rocks which have been tumbled from the cliffs or have been washed in by lateral streams. Immediately above the narrow, rocky channel, on one or both sides, there is often a bay of quiet water, in which a landing can be made with ease. Sometimes the water descends with a smooth, unruffled surface from the broad, quiet spread above into the narrow, angry channel below by a semicircular sag. Great care must be taken not to pass over the brink into this deceptive pit, but above it we can row with safety. I walk along the bank to examine the ground, leaving one of my men with a flag to guide the other boats to the landing-place. I soon see one of the boats make shore all right, and feel no more concern; but a minute after, I hear a shout, and, looking around, see one of the boats shooting down the center of the sag. It is the "No Name," with Captain Howland, his brother, and Goodman. I feel that its going over is inevitable, and run to save the third boat. A minute more, and she turns the point and heads for the shore. Then I turn down stream again and scramble along to look for the boat that has gone over. The first fall is not great, only 10 or 12 feet, and we often run such; but below, the river tumbles down again for 40 or 50 feet, in a channel filled with dangerous rocks that break the waves into whirlpools and beat them into foam. I pass around a great crag just in time to see the boat strike a rock and, rebounding from the shock, careen and fill its open compartment with water. Two of the men lose their oars; she swings around and is carried down at a rapid rate, broadside on, for a few yards, when, striking amidships on another rock with great force, she is broken quite in two and the men are thrown into the river. But the larger part of the boat floats buoyantly, and they soon seize it, and down the river they drift, past the rocks for a few hundred yards, to a second rapid filled with huge boulders, where the boat strikes again and is dashed to pieces, and the men and fragments are soon carried beyond my sight. Running along, I turn a bend and see a man's head above the water, washed about in a whirlpool below a great rock. It is Frank Goodman, clinging to the rock with a grip upon which life depends. Coming opposite, I see Howland trying to go to his aid from an island on which he has been washed. Soon he comes near enough to reach Prank with a pole, which he extends toward him. The latter lets go the rock, grasps the pole, and is pulled ashore. Seneca Howland is washed farther down the island and is caught by some rocks, and, though somewhat bruised, manages to get ashore in safety. This seems a long time as I tell it, but it is quickly done.

And now the three men are on an island, with a swift, dangerous river on either side and a fall below. The "Emma Dean" is soon brought down, and Sumner, starting above as far as possible, pushes out. Right skillfully he plies the oars, and a few strokes set him on the island at the proper point. Then they all pull the boat up stream as far as they are able, until they stand in water up to their necks. One sits on a rock and holds the boat until the others are ready to pull, then gives the boat a push, clings to it with his hands, and climbs in as they pull for mainland, which they reach in safety. We are as glad to shake hands with them as though they had been on a voyage around the world and wrecked on a distant coast.

Down the river half a mile we find that the after cabin of the wrecked boat, with a part of the bottom, ragged and splintered, has floated against a rock and stranded. There are valuable articles in the cabin; but, on examination, we determine that life should not be risked to save them. Of course, the cargo of rations, instruments, and clothing is gone.

We return to the boats and make camp for the night. No sleep comes to me in all those dark hours. The rations, instruments, and clothing have been divided among the boats, anticipating such an accident as this; and we started with duplicates of everything that was deemed necessary to success. But, in the distribution, there was one exception to this precaution–the barometers were all placed in one boat, and they are lost! There is a possibility that they are in the cabin lodged against the rock, for that is where they were kept. But, then, how to reach them? The river is rising. Will they be there to-morrow? Can I go out to Salt Lake City and obtain barometers from New York?

June 10.–I have determined to get the barometers from the wreck, if they are there. After breakfast, while the men make the portage, I go down again for another examination, There the cabin lies, only carried 50 or 60 feet farther on. Carefully looking over the ground, I am satisfied that it can be reached with safety, and return to tell the men my conclusion. Sumner and Dunn volunteer to take the little boat and make the attempt. They start, reach it, and out come the barometers! The boys set up a shout, and I join them, pleased that they should be as glad as myself to save the instruments. When the boat lands on our side, I find that the only things saved from the wreck were the barometers, a package of thermometers, and a three-gallon keg of whiskey. The last is what the men were shouting about. They had taken it aboard unknown to me, and now I am glad they did take it, for it will do them good, as they are drenched every day by the melting snow which runs down from the summits of the Rocky Mountains.

We come back to our work at the portage and find that it is necessary to carry our rations over the rocks for nearly a mile and to let our boats down with lines, except at a few points, where they also must be carried. Between the river and the eastern wall of the canyon there is an immense talus of broken rocks. These have tumbled down from the cliffs above and constitute a vast pile of huge angular fragments. On these we build a path for a quarter of a mile to a small sand-beach covered with driftwood, through which we clear a way for several hundred yards, then continue the trail over another pile of rocks nearly half a mile farther down, to a little bay. The greater part of the day is spent in this work. Then we carry our cargoes down to the beach and camp for the night.

While the men are building the camp fire, we discover an iron bake-oven, several tin plates, a part of a boat, and many other fragments, which denote that this is the place where Ashley's party was wrecked.

June 11.–This day is spent in carrying our rations down to the bay–no small task, climbing over the rocks with sacks of flour and bacon. We carry them by stages of about 500 yards each, and when night comes and the last sack is on the beach, we are tired, bruised, and glad to sleep.

June 12.–To-day we take the boats down to the bay. While at this work we discover three sacks of flour from the wrecked boat that have lodged in the rocks. We carry them above high-water mark and leave them, as our cargoes are already too heavy for the three remaining boats. We also find two or three oars, which we place with them.

As Ashley and his party were wrecked here and as we have lost one of our boats at the same place, we adopt the name Disaster Falls for the scene of so much peril and loss.

Though some of his companions were drowned, Ashley and one other survived the wreck, climbed the canyon wall, and found their way across the Wasatch Mountains to Salt Lake City, living chiefly on berries, as they wandered through an unknown and difficult country. When they arrived at Salt Lake they were almost destitute of clothing and nearly starved. The Mormon people gave them food and clothing and employed them to work on the foundation of the Temple until they had earned sufficient to enable them to leave the country. Of their subsequent history, I have no knowledge. It is possible they returned to the scene of the disaster, as a little creek entering the river below is known as Ashley's Creek, and it is reported that he built a cabin and trapped on this river for one or two winters; but this may have been before the disaster.

June 13.–Rocks, rapids, and portages still. We camp to-night at the foot of the left fall, on a little patch of flood plain covered with a dense growth of box-elders, stopping early in order to spread the clothing and rations to dry. Everything is wet and spoiling.

June 14.–Howland and I climb the wall on the west side of the canyon to an altitude of 2,000 feet. Standing above and looking to the west, we discover a large park, five or six miles wide and twenty or thirty long. The cliff we have climbed forms a wall between the canyon and the park, for it is 800 feet down the western side to the valley. A creek comes winding down 1,200 feet above the river, and, entering the intervening wall by a canyon, plunges down more than 1,000 feet, by a broken cascade, into the river below.

June 15.–To-day, while we make another portage, a peak, standing on the east wall, is climbed by two of the men and found to be 2,700 feet above the river. On the east side of the canyon a vast amphitheater has been cut, with massive buttresses and deep, dark alcoves in which grow beautiful mosses and delicate ferns, while springs burst out from the farther recesses and wind in silver threads over floors of sand rock. Here we have three falls in close succession. At the first the wa$er is compressed into a very narrow channel against the right-hand cliff, and falls 15 feet in 10 yards. At the second we have a broad sheet of water tumbling down 20 feet over a group of rocks that thrust their dark heads through the foam. The third is a broken fall, or short, abrupt rapid, where the water makes a descent of more than 20 feet among huge, fallen fragments of the cliff. We name the group Triplet Falls. We make a portage around the first; past the second and the third we let down with lines.

During the afternoon, Dunn and Howland having returned from their climb, we run down three quarters of a mile on quiet waters and land at the head of another fall. On examination, we find that there is an abrupt plunge of a few feet and then the river tumbles for half a mile with a descent of a hundred feet, in a channel beset with great numbers of huge boulders. This stretch of the river is named Hell's Half-Mile. The remaining portion of the day is occupied in making a trail among the rocks at the foot of the rapid.

June 16.–Our first work this morning is to carry our cargoes to the foot of the falls. Then we commence letting down the boats. We take two of them down in safety, but not without great difficulty; for, where such a vast body of water, rolling down an inclined plane, is broken into eddies and cross-currents by rocks projecting from the cliffs and piles of boulders in the channel, it requires excessive labor and much care to prevent the boats from being dashed against the rocks or breaking away. Sometimes we are compelled to hold the boat against a rock above a chute until a second line, attached to the stem, is carried to some point below, and when all is ready the first line is detached and the boat given to the current, when she shoots down and the men below swing her into some eddy.

At such a place we are letting down the last boat, and as she is set free a wave turns her broadside down the stream, with the stem, to which the line is attached, from shore and a little up. They haul on the line to bring the boat in, but the power of the current, striking obliquely against her, shoots her out into the middle of the river. The men have their hands burned with the friction of the passing line; the boat breaks away and speeds with great velocity down the stream. The "Maid of the Canyon" is lost! So it seems; but she drifts some distance and swings into an eddy, in which she spins about until we arrive with the small boat and rescue her.

Soon we are on our way again, and stop at the mouth of a little brook on the right for a late dinner. This brook comes down from the distant mountains in a deep side canyon. We set out to explore it, but are soon cut off from farther progress up the gorge by a high rock, over which the brook glides in a smooth sheet. The rock is not quite vertical, and the water does not plunge over it in a fall.

Then we climb up to the left for an hour, and are 1,000 feet above the river and 600 above the brook. Just before us the canyon divides, a little stream coming down on the right and another on the left, and we can look away up either of these canyons, through an ascending vista, to cliffs and crags and towers a mile back and 2,000 feet overhead. To the right a dozen gleaming cascades are seen. Pines and firs stand on the rocks and aspens overhang the brooks. The rocks below are red and brown, set in deep shadows, but above they are buff and vermilion and stand in the sunshine. The light above, made more brilliant by the bright-tinted rocks, and the shadows below, more gloomy by reason of the somber hues of the brown walls, increase the apparent depths of the canyons, and it seems a long way up to the world of sunshine and open sky, and a long way down to the bottom of the canyon glooms. Never before have I received such an impression of the vast heights of these canyon walls, not even at the Cliff of the Harp, where the very heavens seemed to rest on their summits. We sit on some overhanging rocks and enjoy the scene for a time, listening to the music of the falling waters away up the canyon. We name this Rippling Brook.

Late in the afternoon we make a short run to the mouth of another little creek, coming down from the left into an alcove filled with luxuriant vegetation. Here camp is made, with a group of cedars on one side and a dense mass of box-elders and dead willows on the other.

I go up to explore the alcove. While away a whirlwind comes and scatters the fire among the dead willows and cedar-spray, and soon there is a conflagration. The men rush for the boats, leaving all they cannot readily seize at the moment, and even then they have their clothing burned and hair singed, and Bradley has his ears scorched. The cook fills his arms with the mess-kit, and jumping into a boat, stumbles and falls, and away go our cooking utensils into the river. Our plates are gone; our spoons are gone; our knives and forks are gone. "Water catch 'em; h-e-a-p catch 'em."

When on the boats, the men are compelled to cut loose, as the flames, running out on the overhanging willows, are scorching them. Loose on the stream, they must go down, for the water is too swift to make headway against it. Just below is a rapid, filled with rocks. On the shoot, no channel explored, no signal to guide them! Just at this juncture I chance to see them, but have not yet discovered the fire, and the strange movements of the men fill me with astonishment. Down the rocks I clamber, and run to the bank. When I arrive they have landed. Then we all go back to the late camp to see if anything left behind can be saved. Some of the clothing and bedding taken out of the boats is found, also a few tin cups, basins, and a camp kettle; and this is all the mess-kit we now have. Yet we do just as well as ever.

June 17.–We run down to the mouth of Yampa River. This has been a chapter of disasters and toils, notwithstanding which the Canyon of Lodore was not devoid of scenic interest, even beyond the power of pen to tell. The roar of its waters was heard unceasingly from the hour we entered it until we landed here. No quiet in all that time. But its walls and cliffs, its peaks and crags, its amphitheaters and alcoves, tell a story of beauty and grandeur that I hear yet–and shall hear.

The Canyon of Lodore is 20 3/4 miles in length. It starts abruptly at what we have called the Gate of Lodore, with walls nearly 2,000 feet high, and they are never lower than this until we reach Alcove

Brook, about three miles above the foot. They are very irregular, standing in vertical or overhanging cliffs in places, terraced in others, or receding in steep slopes, and are broken by many side gulches and canyons. The highest point on the wall is at Dunn's Cliff, near Triplet Falls, where the rocks reach an altitude of 2,700 feet, but the peaks a little way back rise nearly 1,000 feet higher. Yellow pines, nut pines, firs, and cedars stand in extensive forests on the Uinta Mountains, and, clinging to the rocks and growing in the crevices, come down the walls to the water's edge from Flaming Gorge to Echo Park. The red sandstones are lichened over; delicate mosses grow in the moist places, and ferns festoon the walls.

CHAPTER VIII. FROM ECHO PARK TO THE MOUTH OF THE UINTA RIVER.

The Yampa enters the Green from the east. At a point opposite its mouth the Green runs to the south, at the foot of a rock about 700 feet high and a mile long, and then turns sharply around the rock to the right and runs back in a northerly course parallel to its former direction for nearly another mile, thus having the opposite sides of a long, narrow rock for its bank. The tongue of rock so formed is a peninsular precipice with a mural escarpment along its whole course on the east, but broken down at places on the west.

On the east side of the river, opposite the rock and below the Yampa, there is a little park, just large enough for a farm, already fenced with high walls of gray homogeneous sandstone. There are three river entrances to this park: one down the Yampa; one below, by coming up the Green; and another down the Green. There is also a land entrance down a lateral canyon. Elsewhere the park is inaccessible. Through this land entrance by the side canyon there is a trail made by Indian hunters, who come down here in certain seasons to kill mountain sheep. Great hollow domes are seen in the eastern side of the rock, against which the Green sweeps; willows border the river; clumps of box-elder are seen; and a few cottonwoods stand at the lower end. Standing opposite the rock, our words are repeated with startling clearness, but in a soft, mellow tone, that transforms them into magical music. Scarcely can one believe it is the echo of his own voice. In some places two or three echoes come back; in other places they repeat themselves, passing back and forth across the river between this rock and the eastern wall. To hear these repeated echoes well, we must shout. Some of the party aver that ten or twelve repetitions can be heard. To me, they seem rapidly to diminish and merge by multiplicity, like telegraph poles on an outstretched plain. I have observed the same phenomenon once before in the cliffs near Long's Peak, and am pleased to meet with it again.

During the afternoon Bradley and I climb some cliffs to the north. Mountain sheep are seen above us, and they stand out on the rocks and eye us intently, not seeming to move. Their color is much like that of the gray sandstone beneath them, and, immovable as they are, they appear like carved forms. Now a fine ram beats the rock with his fore foot, and, wheeling around, they all bound away together, leaping over rocks and chasms and climbing walls where no man can follow, and this with an ease and grace most wonderful. At night we return to our camp under the box-elders by the river side. Here we are to spend two or three days, making a series of astronomic observations for latitude and longitude.

June 18.–We have named the long peninsular rock on the other side Echo Rock. Desiring to climb it, Bradley and I take the little boat and pull up stream as far as possible, for it cannot be climbed directly opposite. We land on a talus of rocks at the upper end in order to reach a place where it seems practicable to make the ascent; but we find we must go still farther up the river. So we scramble along, until we reach a place where the river sweeps against the wall. Here we find a shelf along which we can pass, and now are ready for the climb.

We start up a gulch; then pass to the left on a bench along the wall; then up again over broken rocks; then we reach more benches, along which we walk, until we find more broken rocks and crevices, by which we climb; still up, until we have ascended 600 or 800 feet, when we are met by a sheer precipice. Looking about, we find a place where it seems possible to climb. I go ahead; Bradley hands the barometer to me, and follows. So we proceed, stage by stage, until we are nearly to the summit. Here, by making a spring, I gain a foothold in a little crevice, and grasp an angle of the rock overhead. I find I can get up no farther and cannot step back, for I dare not let go with my hand and cannot reach foothold below without. I call to Bradley for help. He finds a way by which he can get to the top of the rock over my head, but cannot reach me. Then he looks around for some stick or limb of a tree, but finds none. Then he suggests that he would better help me with the barometer case, but I fear I cannot hold on to it. The moment is critical. Standing on my toes, my muscles begin to tremble. It is sixty or eighty feet to the foot of the precipice. If I lose my hold I shall fall to the bottom and then perhaps roll over the bench and tumble still farther down the cliff. At this instant it occurs to Bradley to take off his drawers, which he does, and swings them down to me. I hug close to the rock, let go with my hand, seize the dangling legs, and with his assistance am enabled to gain the top.

Then we walk out on the peninsular rock, make the necessary observations for determining its altitude above camp, and return, finding an easy way down.

June 19.–To-day, Howland, Bradley, and I take the "Emma Dean" and start up the Yampa River. The stream is much swollen, the current swift, and we are able to make but slow progress against it. The canyon in this part of the course of the Yampa is cut through light gray sandstone. The river is very winding, and the swifter water is usually found on the outside of the curve, sweeping against vertical cliffs often a thousand feet high. In the center of these curves, in many places, the rock above overhangs the river. On the opposite side the walls are broken, craggy, and sloping, and occasionally side canyons enter. When we have rowed until we are quite tired we stop and take advantage of one of these broken places to climb out of the canyon. When above, we can look up the Yampa for a distance of several miles. From the summit of the immediate walls of the canyon the rocks rise gently back for a distance of a mile or two, having the appearance of a valley with an irregular and rounded sandstone floor and in the center a deep gorge, which is the canyon. The rim of this valley on the north is from 2,500 to 3,000 feet above the river; on the south it is not so high. A number of peaks stand on this northern rim, the highest of which has received the name Mount Dawes.

Late in the afternoon we descend to our boat and return to camp in Echo Park, gliding down in twenty minutes on the rapid river, a distance of four or five miles, which was made up stream only by several hours' hard rowing in the morning.

June 20.–This morning two of the men take me up the Yampa for a short distance, and I go out to climb. Having reached the top of the canyon, I walk over long stretches of naked sandstone, crossing gulches now and then, and by noon reach the summit of Mount

Dawes. From this point I can look away to the north and see in the dim distance the Sweetwater and Wind River mountains, more than 100 miles away. To the northwest the Wasatch Mountains are in view, and peaks of the Uinta. To the east I can see the western slopes of the Rocky Mountains, more than 150 miles distant. The air is singularly clear to-day; mountains and buttes stand in sharp outline, valleys stretch out in perspective, and I can look down into the deep canyon gorges and see gleaming waters.

Descending, I cross to a ridge near the brink of the Canyon of Lodore, the highest point of which is nearly as high as the last mentioned mountain. Late in the afternoon I stand on this elevated point and discover a monument that has evidently been built by human hands. A few plants are growing in the joints between the rocks, and all are lichened over to a greater or less extent, giving evidence that the pile was built a long time ago. This line of peaks, the eastern extension of the Uinta Mountains, has received the name of Sierra Escalante, in honor of a Spanish priest who traveled in this region of country nearly a century ago. Perchance the reverend father built this monument.

Now I return to the river and discharge my gun, as a signal for the boat to come and take me down to camp. While we have been in the park the men have succeeded in catching a number of fish, and we have an abundant supply. This is a delightful addition to our menu.

June 21.– We float around the long rock and enter another canyon. The walls are high and vertical, the canyon is narrow, and the river fills the whole space below, so that there is no landing-place at the foot of the cliff. The Green is greatly increased by the Yampa, and we now have a much larger river. All this volume of water, confined, as it is, in a narrow channel and rushing with great velocity, is set eddying and spinning in whirlpools by projecting rocks and short curves, and the waters waltz their way through the canyon, making their own rippling, rushing, roaring music. The canyon is much narrower than any we have seen. We manage our boats with difficulty. They spin about from side to side and we know not where we are going, and find it impossible to keep them headed down the stream. At first this causes us great alarm, but we soon find there is little danger, and that there is a general movement or progression down the river, to which this whirling is but an adjunct–that it is the merry mood of the river to dance through this deep, dark gorge, and right gaily do we join in the sport.

But soon our revel is interrupted by a cataract; its roaring command is heeded by all our power at the oars, and we pull against the whirling current. The "Emma Dean" is brought up against a cliff about 50 feet above the brink of the fall. By vigorously plying the oars on the side opposite the wall, as if to pull up stream, we can hold her against the rock. The boats behind are signaled to land where they can. The "Maid of the Canyon" is pulled to the left wall, and, by constant rowing, they can hold her also. The "Sister" is run into an alcove on the right, where an eddy is in a dance, and in this she joins. Now my little boat is held against the wall only by the utmost exertion, and it is impossible to make headway against the current. On examination, I find a horizontal crevice in the rock, about 10 feet above the water and a boat's length below us; so we let her down to that point. One of the men clambers into the crevice, into which he can just crawl; we toss him the line, which he makes fast in the rocks, and now our boat is tied up. Then I follow into the crevice and we crawl along up stream a distance of 50 feet or more, and find a broken place where we can climb about 50 feet higher. Here we stand on a shelf that passes along down stream to a point above the falls, where it is broken down, and a pile of rocks, over which we can descend to the river, is lying against the foot of the cliff.

It has been mentioned that one of the boats is on the other side. I signal for the men to pull her up alongside of the wall, but it cannot be done; then to cross. This they do, gaining the wall on our side just above where the "Emma Dean" is tied.

The third boat is out of sight, whirling in the eddy of a recess. Looking about, I find another horizontal crevice, along which I crawl to a point just over the water where this boat is lying, and, calling loud and long, I finally succeed in making the crew understand that I want them to bring the boat down, hugging the wall. This they accomplish by taking advantage of every crevice and knob on the face of the cliff, so that we have the three boats together at a point a few yards above the falls. Now, by passing a line up on the shelf, the boats can be let down to the broken rocks below. This we do, and, making a short portage, our troubles here are over.

Below the falls the canyon is wider, and there is more or less space between the river and the walls; but the stream, though wide, is rapid, and rolls at a fearful rate among the rocks. We proceed with great caution, and run the large boats wholly by signal.

At night we camp at the mouth of a small creek, which affords us a good supper of trout. In camp to-night we discuss the propriety of several different names for this canyon. At the falls encountered at noon its characteristics change suddenly. Above, it is very narrow, and the walls are almost vertical; below, the canyon is much wider and more flaring, and high up on the sides crags, pinnacles, and towers are seen. A number of wild and narrow side canyons enter, and the walls are much broken. After many suggestions our choice rests between two names, Whirlpool Canyon and Craggy Canyon, neither of which is strictly appropriate for both parts of it; so we leave the discussion at this point, with the understanding that it is best, before finally deciding on a name, to wait until we see what the character of the canyon is below.

June 22.–Still making short portages and letting down with lines. While we are waiting for dinner to-day, I climb a point that gives me a good view of the river for two or three miles below, and I think we can make a long run. After dinner we start; the large boats are to follow in fifteen minutes and look out for the signal to land. Into the middle of the stream we row, and down the rapid river we glide, only making strokes enough with the oars to guide the boat. What a headlong ride it is! shooting past rocks and islands. I am soon filled with exhilaration only experienced before in riding a fleet horse over the outstretched prairie. One, two, three, four miles we go, rearing and plunging with the waves, until we wheel to the right into a beautiful park and land on an island, where we go into camp.

An hour or two before sunset I cross to the mainland and climb a point of rocks where I can overlook the park and its surroundings.

On the east it is bounded by a high mountain ridge. A semicircle of naked hills bounds it on the north, west, and south.

The broad, deep river meanders through the park, interrupted by many wooded islands; so I name it Island Park, and decide to call the canyon above, Whirlpool Canyon.

June 23.–We remain in camp to-day to repair our boats, which have had hard knocks and are leaking. Two of the men go out with the barometer to climb the cliff at the foot of Whirlpool Canyon and measure the walls; another goes on the mountain to hunt; and Bradley and I spend the day among the rocks, studying an interesting geologic fold and collecting fossils. Late in the afternoon the hunter returns and brings with him a fine, fat deer; so we give his name to the mountain–Mount Hawkins. Just before night we move camp to the lower end of the park, floating down the river about four miles.

June 24.–Bradley and I start early to climb the mountain ridge to the east, and find its summit to be nearly 3,000 feet above camp. It has required some labor to scale it; but on its top, what a view! There is a long spur running out from the Uinta Mountains toward the south, and the river runs lengthwise through it. Coming down Lodore and Whirlpool canyons, we cut through the southern slope of the Uinta Mountains; and the lower end of this latter canyon runs into the spur, but, instead of splitting it the whole length, the river wheels to the right at the foot of Whirlpool Canyon in a great curve to the northwest through Island Park. At the lower end of the park, the river turns again to the southeast and cuts into the mountain to its center and then makes a detour to the southwest, splitting the mountain ridge for a distance of six miles nearly to its foot, and then turns out of it to the left. All this we can see where we stand on the summit of Mount Hawkins, and so we name the gorge below, Split Mountain Canyon.

We are standing 3,000 feet above the waters, which are troubled with billows and are white with foam. The walls are set with crags and peaks and buttressed towers and overhanging domes. Turning to the right, the park is below us, its island groves reflected by the deep, quiet waters. Rich meadows stretch out on either hand to the verge of a sloping plain that comes down from the distant mountains. These plains are of almost naked rock, in strange contrast to the meadows,–blue and lilac colored rocks, buff and pink, vermilion and brown, and all these colors clear and bright. A dozen little creeks, dry the greater part of the year, run down through the half circle of exposed formations, radiating from the island center to the rim of the basin. Each creek has its system of side streams and each side stream has its system of laterals, and again these are divided; so that this outstretched slope of rock is elaborately embossed. Beds of different-colored formations run in parallel bands on either side. The perspective, modified by the undulations, gives the bands a waved appearance, and the high colors gleam in the midday sun with the luster of satin. We are tempted to call this Rainbow Park. Away beyond these beds are the Uinta and Wasatch mountains with their pine forests and snow fields and naked peaks. Now we turn to the right and look up Whirlpool Canyon, a deep gorge with a river at the bottom–a gloomy chasm, where mad waves roar; but at this distance and altitude the river is but a rippling brook, and the chasm a narrow cleft. The top of the mountain on which we stand is a broad, grassy table, and a herd of deer are feeding in the distance. Walking over to the southeast, we look down into the valley of White River, and beyond that see the far-distant Rocky Mountains, in mellow, perspective haze, through which snow fields shine.

June 25.–This morning we enter Split Mountain Canyon, sailing in through a broad, flaring, brilliant gateway. We run two or three rapids, after they have been carefully examined. Then we have a series of six or eight, over which we are compelled to pass by letting the boats down with lines. This occupies the entire day, and we camp at night at the mouth of a great cave. The cave is at the foot of one of these rapids, and the waves dash in nearly to its end. We can pass along a little shelf at the side until we reach the back part. Swallows have built their nests in the ceiling, and they wheel in, chattering and scolding at our intrusion; but their clamor is almost drowned by the noise of the waters. Looking out of the cave, we can see, far up the river, a line of crags standing sentinel on either side, and Mount Hawkins in the distance.

June 26.–The forenoon is spent in getting our large boats over the rapids. This afternoon we find three falls in close succession. We carry our rations over the rocks and let our boats shoot over the falls, checking and bringing them to land with lines in the eddies below. At three o'clock we are all aboard again. Down the river we are carried by the swift waters at great speed, sheering around a rock now and then with a timely stroke or two of the oars. At one point the river turns from left to right, in a direction at right angles to the canyon, in a long chute and strikes the right, where its waters are heaped up in great billows that tumble back in breakers. We glide into the chute before we see the danger, and it is too late to stop. Two or three hard strokes are given on the right and we pause for an instant, expecting to be dashed against the rock. But the bow of the boat leaps high on a great wave, the rebounding waters hurl us back, and the peril is past. The next moment the other boats are hurriedly signaled to land on the left. Accomplishing this, the men walk along the shore, holding the boats near the bank, and let them drift around. Starting again, we soon debouch into a beautiful valley, glide down its length for 10 miles, and camp under a grand old cottonwood. This is evidently a frequent resort for Indians. Tent poles are lying about, and the dead embers of late camp fires are seen. On the plains to the left, antelope are feeding. Now and then a wolf is seen, and after dark they make the air resound with their howling.

June 27.–Now our way is along a gently flowing river, beset with many islands; groves are seen on either side, and natural meadows, where herds of antelope are feeding. Here and there we have views of the distant mountains on the right. During the afternoon we make a long detour to the west and return again to a point not more than half a mile from where we started at noon, and here we camp for the night under a high bluff. June 28.–To-day the scenery on either side of the river is much the same as that of yesterday, except that two or three lakes are discovered, lying in the valley to the west. After dinner we run but a few minutes when we discover the mouth of the Uinta, a river coming in from the west. Up the valley of this stream about 40 miles the reservation of the Uinta Indians is situated. We propose to go there and see if we can replenish our mess-kit, and perhaps send letters to friends. We also desire to establish an astronomic station here; and hence this will be our stopping place for several days.

Some years ago Captain Berthoud surveyed a stage route from Salt Lake City to Denver, and this is the place where he crossed the Green River. His party was encamped here for some time, constructing a ferry boat and opening a road.

A little above the mouth of the Uinta, on the west side of the Green, there is a lake of several thousand acres. We carry our boat across the divide between this and the river, have a row on its quiet waters, and succeed in shooting several ducks.

June 29.–A mile and three quarters from here is the junction of the White River with the Green. The White has its source far to the east in the Rocky Mountains. This morning I cross the Green and go over into the valley of the White and extend my walk several miles along its winding way, until at last I come in sight of some strangely carved rocks, named by General Hughes, in his journal, "Goblin City." Our last winter's camp was situated a hundred miles above the point reached to-day. The course of the river, for much of the distance, is through canyons; but at some places valleys are found. Excepting these little valleys, the region is one of great desolation: arid, almost treeless, with bluffs, hills, ledges of rock, and drifting sands. Along the course of the Green, however, from the foot of Split Mountain Canyon to a point some distance below the mouth of the Uinta, there are many groves of cottonwood, natural meadows, and rich lands. This arable belt extends some distance up the White River on the east and the Uinta on the west, and the time must soon come when settlers will penetrate this country and make homes.

June 30.–We have a row up the Uinta to-day, but are not able to make much headway against the swift current, and hence conclude we must walk all the way to the agency.

July 1.–Two days have been employed in obtaining the local time, taking observations for latitude and longitude, and making excursions into the adjacent country. This morning, with two of the men, I start for the agency. It is a toilsome walk, 20 miles of the distance being across a sand desert. Occasionally we have to wade the river, crossing it back and forth. Toward evening we cross several beautiful streams, tributaries of the Uinta, and pass through pine groves and meadows, arriving at the reservation just at dusk. Captain Dodds, the agent, is away, having gone to Salt Lake City, but his assistants receive us very kindly. It is rather pleasant to see a house once more, and some evidences of civilization, even if it is on an Indian reservation several days' ride from the nearest home of the white man.

July 2.–I go this morning to visit Tsauwiat. This old chief is but the wreck of a man, and no longer has influence. Looking at him one can scarcely realize that he is a man. His skin is shrunken, wrinkled, and dry, and seems to cover no more than a form of bones. He is said to be more than 100 years old. I talk a little with him, but his conversation is incoherent, though he seems to take pride in showing me some medals that must have been given him many years ago. He has a pipe which he says he has used a long time. I offer to exchange with him, and he seems to be glad to accept; so I add another to my collection of pipes. His wife, "The Bishop," as she is called, is a very garrulous old woman; she exerts a great influence, and is much revered. She is the only Indian woman I have known to occupy a place in the council ring. She seems very much younger than her husband, and, though wrinkled and ugly, is still vigorous. She has much to say to me concerning the condition of the people, and seems very anxious that they should learn to cultivate the soil, own farms, and live like white men. After talking a couple of hours with these old people, I go to see the farms. They are situated in a very beautiful district, where many fine streams of water meander across alluvial plains and meadows. These creeks have a considerable fall, and it is easy to take their waters out above and overflow the lands with them.

It will be remembered that irrigation is necessary in this dry climate to successful farming. Quite a number of Indians have each a patch of ground of two or three acres, on which they are raising wheat, potatoes, turnips, pumpkins, melons, and other vegetables. Most of the crops are looking well, and it is rather surprising with what pride they show us that they are able to cultivate crops like white men. They are still occupying lodges, and refuse to build houses, assigning as a reason that when any one dies in a lodge it is always abandoned, and very often burned with all the effects of the deceased; and when houses have been built for them the houses have been treated in the same way. With their unclean habits, a fixed residence would doubtless be no pleasant place.

This beautiful valley has been the home of a people of a higher grade of civilization than the present Utes. Evidences of this are quite abundant; on our way here yesterday we discovered fragments of pottery in many places along the trail; and, wandering about the little farms to-day, I find the foundations of ancient houses, and mealing-stones that were not used by nomadic people, as they are too heavy to be transported by such tribes, and are deeply worn. The Indians, seeing that I am interested in these matters, take pains to show me several other places where these evidences remain, and tell me that they know nothing about the people who formerly dwelt here. They further tell me that up in the canyon the rocks are covered with pictures.

July 5.–The last two days have been spent in studying the language of the Indians and in making collections of articles illustrating the state of arts among them.

Frank Goodman informs me this morning that he has concluded not to go on with the party, saying that he has seen danger enough. It will be remembered that he was one of the crew on the "No Name" when she was wrecked. As our boats are rather heavily loaded, I am content that he should leave, although he has been a faithful man.

We start early on our return to the boats, taking horses with us from the reservation, and two Indians, who are to bring the animals back.

Whirlpool Canyon is 14 1/4 miles in length, the walls varying from 1,800 to 2,400 feet in height. The course of the river through Island Park is 9 miles. Split Mountain Canyon is 8 miles long. The highest crags on its walls reach an altitude above the river of from 2,500 to 2,700 feet. In these canyons cedars only are found on the walls.

The distance by river from the foot of Split Mountain Canyon to the mouth of the Uinta is 67 miles. The valley through which it runs is the home of many antelope, and we have adopted for it the Indian name Won'sits Yuav–Antelope Valley.

CHAPTER IX. FROM THE MOUTH OF THE UINTA RIVER TO THE JUNCTION OF THE GRAND AND GREEN.

July 6.–An early start this morning. A short distance below the mouth of the Uinta we come to the head of a long island. Last winter a man named Johnson, a hunter and Indian trader, visited us at our camp in White River Valley. This man has an Indian wife, and, having no fixed home, usually travels with one of the Ute bands. He informed me that it was his intention to plant some corn, potatoes, and other vegetables on this island in the spring, and, knowing that we would pass it, invited us to stop and help ourselves, even if he should not be there; so we land and go out on the island. Looking about, we soon discover his garden, but it is in a sad condition, having received no care since it was planted. It is yet too early in the season for corn, but Hall suggests that potato tops are good greens, and, anxious for some change from our salt-meat fare, we gather a quantity and take them aboard. At noon we stop and cook our greens for dinner; but soon one after another of the party is taken sick; nausea first, and then severe vomiting, and we tumble around under the trees, groaning with pain. I feel a little alarmed, lest our poisoning be severe. Emetics are administered to those who are willing to take them, and about the middle of the afternoon we are all rid of the pain. Jack Sumner records in his diary that "potato tops are not good greens on the 6th day of July."

This evening we enter another canyon, almost imperceptibly, as the walls rise very gently.

July 7.–We find quiet water to-day, the river sweeping in great and beautiful curves, the canyon walls steadily increasing in altitude. The escarpments formed by the cut edges of the rock are often vertical, sometimes terraced, and in some places the treads of the terraces are sloping. In these quiet curves vast amphitheaters are formed, now in vertical rocks, now in steps.

The salient point of rock within the curve is usually broken down in a steep slope, and we stop occasionally to climb up at such a place, where on looking down we can see the river sweeping the foot of the opposite cliff in a great, easy curve, with a perpendicular or terraced wall rising from the water's edge many hundreds of feet. One of these we find very symmetrical and name it Sumner's Amphitheater. The cliffs are rarely broken by the entrance of side canyons, and we sweep around curve after curve with almost continuous walls for several miles.

Late in the afternoon we find the river very much rougher and come upon rapids, not dangerous, but still demanding close attention. We camp at night on the right bank, having made 26 miles. July 8.–This morning Bradley and I go out to climb, and gain an altitude of more than 2,000 feet above the river, but still do not reach the summit of the wall.

After dinner we pass through a region of the wildest desolation. The canyon is very tortuous, the river very rapid, and many lateral canyons enter on either side. These usually have their branches, so that the region is cut into a wilderness of gray and brown cliffs. In several places these lateral canyons are separated from one another only by narrow walls, often hundreds of feet high,–so narrow in places that where softer rocks are found below they have crumbled away and left holes in the wall, forming passages from one canyon into another. These we often call natural bridges; but they were never intended to span streams. They would better, perhaps, be called side doors between canyon chambers. Piles of broken rock lie against these walls; crags and tower-shaped peaks are seen everywhere, and away above them, long lines of broken cliffs; and above and beyond the cliffs are pine forests, of which we obtain occasional glimpses as we look up through a vista of rocks. The walls are almost without vegetation; a few dwarf bushes are seen here and there clinging to the rocks, and cedars grow from the crevices–not like the cedars of a land refreshed with rains, great cones bedecked with spray, but ugly clumps, like war clubs beset with spines. We are minded to call this the Canyon of Desolation.

The wind annoys us much to-day. The water, rough by reason of the rapids, is made more so by head gales. Wherever a great face of rocks has a southern exposure, the rarefied air rises and the wind rushes in below, either up or down the canyon, or both, causing local currents. Just at sunset we run a bad rapid and camp at its foot.

July 9.–Our run to-day is through a canyon with ragged, broken walls, many lateral gulches or canyons entering on either side. The river is rough, and occasionally it becomes necessary to use lines in passing rocky places. During the afternoon we come to a rather open canyon valley, stretching up toward the west, its farther end lost in the mountains. From a point to which we climb we obtain a good view of its course, until its angular walls are lost in the vista.

July 10.–Sumner, who is a fine mechanic, is learning to take observations for time with the sextant. To-day he remains in camp to practice. Howland and I determine to climb out, and start up a lateral canyon, taking a barometer with us for the purpose of measuring the thickness of the strata over which we pass. The readings of the barometer below are recorded every half hour and our observations must be simultaneous. Where the beds which we desire to measure are very thick, we must climb with the utmost speed to reach their summits in time; where the beds are thinner, we must wait for the moment to arrive; and so, by hard and easy stages, we make our way to the top of the canyon wall and reach the plateau above about two o' clock.

Howland, who has his gun with him, sees deer feeding a mile or two back and goes off for a hunt. I go to a peak which seems to be the highest one in this region, about half a mile distant, and climb, for-the purpose of tracing the topography of the adjacent country. From this point a fine view is obtained. A long plateau stretches across the river in an easterly and westerly direction, the summit covered by pine forests, with intervening elevated valleys and gulches. The plateau itself is cut in two by the canyon. Other side canyons head away back from the river and run down into the Green. Besides these, deep and abrupt canyons are seen to head back on the plateau and run north toward the Uinta and White rivers. Still other canyons head in the valleys and run toward the south. The elevation of the plateau being about 8,000 feet above the level of the sea, it is in a region of moisture, as is well attested by the forests and grassy valleys. The plateau seems to rise gradually to the west, until it merges into the Wasatch Mountains. On these high table-lands elk and deer abound; and they are favorite hunting grounds for the Ute Indians.

A little before sunset Howland and I meet again at the head of the side canyon, and down we start. It is late, and we must make great haste or be caught by the darkness; so we go, running where we can, leaping over the ledges, letting each other down on the loose rocks, as long as we can see. When darkness comes we are still some distance from camp, and a long, slow, anxious descent is made toward the gleaming camp fire.

After supper, observations for latitude are taken, and only two or three hours for sleep remain before daylight.

July 11.– A short distance below camp we run a rapid, and in doing so break an oar and then lose another, both belonging to the "Emma Dean." Now the pioneer boat has but two oars. We see nothing from which oars can be made, so we conclude to run on to some point where it seems possible to climb out to the forests on the plateau, and there we will procure suitable timber from which to make new ones.

We soon approach another rapid. Standing on deck, I think it can be run, and on we go. Coming nearer, I see that at the foot it has a short turn to the left, where the waters pile up against the cliff. Here we try to land, but quickly discover that, being in swift water above the fall, we cannot reach shore, crippled as we are by the loss of two oars; so the bow of the boat is turned down stream. We shoot by a big rock; a reflex wave rolls over our little boat and fills her. I see that the place is dangerous and quickly signal to the other boats to land where they can. This is scarcely completed when another wave rolls our boat over and I am thrown some distance into the water. I soon find that swimming is very easy and I cannot sink. It is only necessary to ply strokes sufficient to keep my head out of the water, though now and then, when a breaker rolls over me, I close my mouth and am carried through it. The boat is drifting ahead of me 20 or 30 feet, and when the great waves have passed I overtake her and find Sumner and Dunn clinging to her. As soon as we reach quiet water we all swim to one side and turn her over. In doing this, Dunn loses his hold and goes under; when he comes up he is caught by Sumner and pulled to the boat. In the meantime we have drifted down stream some distance and see another rapid below. How bad it may be we cannot tell; so we swim toward shore, pulling our boat with us, with all the vigor possible, but are carried down much faster than distance toward shore is diminished. At last we reach a huge pile of driftwood. Our rolls of blankets, two guns, and a barometer were in the open compartment of the boat and, when it went over, these were thrown out. The guns and barometer are lost, but I succeeded in catching one of the rolls of blankets as it drifted down, when we were swimming to shore; the other two are lost, and sometimes hereafter we may sleep cold.

A huge fire is built on the bank and our clothing spread to dry, and then from the drift logs we select one from which we think oars can be made, and the remainder of the day is spent in sawing them out.

July 12.–This morning the new oars are finished and we start once more. We pass several bad rapids, making a short portage at one, and before noon we come to a long, bad fall, where the channel is filled with rocks on the left which turn the waters to the right, where they pass under an overhanging rock. On examination we determine to run it, keeping as close to the left-hand rocks as safety will permit, in order to avoid the overhanging cliff. The little boat runs over all right; another follows, but the men are not able to keep her near enough to the left bank and she is carried by a swift chute into great waves to the right, where she is tossed about and Bradley is knocked over the side; his foot catching under the seat, he is dragged along in the water with his head down; making great exertion, he seizes the gunwale with his left hand and can lift his head above water now and then. To us who are below, it seems impossible to keep the boat from going under the overhanging cliff; but Powell, for the moment heedless of Bradley's mishap, pulls with all his power for half a dozen strokes, when the danger is past; then he seizes Bradley and pulls him in. The men in the boat above, seeing this, land, and she is let down by lines.

Just here we emerge from the Canyon of Desolation, as we have named it, into a more open country, which extends for a distance of nearly a mile, when we enter another canyon cut through gray sandstone.

About three o'clock in the afternoon we meet with a new difficulty. The river fills the entire channel; the walls are vertical on either side from the water's edge, and a bad rapid is beset with rocks. We come to the head of it and land on a rock in the stream. The little boat is let down to another rock below, the men of the larger boat holding to the line; the second boat is let down in the same way, and the line of the third boat is brought with them. Now the third boat pushes out from the upper rock, and, as we have her line below, we pull in and catch her as she is sweeping by at the foot of the rock on which we stand. Again the first boat is let down stream the full length of her line and the second boat is passed down, by the first to the extent of her line, which is held by the men in the first boat; so she is two lines' length from where she started. Then the third boat is let down past the second, and still down, nearly to the length of her line, so that she is fast to the second boat and swinging down three lines' lengths, with the other two boats intervening. Held in this way, the men are able to pull her into a cove in the left wall, where she is made fast. But this leaves a man on the rock above, holding to the line of the little boat. When all is ready, he springs from the rock, clinging to the line with one hand and swimming with the other, and we pull him in as he goes by. As the two boats, thus loosened, drift down, the men in the cove pull us all in as we come opposite; then we pass around to a point of rock below the cove, close to the wall, land, make a short portage over the worst places in the rapid, and start again.

At night we camp on a sand beach. The wind blows a hurricane; the drifting sand almost blinds us; and nowhere can we find shelter. The wind continues to blow all night, the sand sifting through our blankets and piling over us until we are covered as in a snowdrift. We are glad when morning comes.

July 13.–This morning we have an exhilarating ride. The river is swift, and there are many smooth rapids. I stand on deck, keeping careful watch ahead, and we glide along, mile after mile, plying strokes, now on the right and then on the left, just sufficient to guide our boats past the rocks into smooth water. At noon we emerge from Gray Canyon, as we have named it, and camp for dinner under a cotton-wood tree standing on the left bank.

Extensive sand plains extend back from the immediate river valley as far as we can see on either side. These naked, drifting sands gleam brilliantly in the midday sun of July. The reflected heat from the glaring surface produces a curious motion of the atmosphere; little currents are generated and the whole seems to be trembling and moving about in many directions, or, failing to see that the movement is in the atmosphere, it gives the impression of an unstable land. Plains and hills and cliffs and distant mountains seem to be floating vaguely about in a trembling, wave-rocked sea, and patches of landscape seem to float away and be lost, and then to reappear.

Just opposite, there are buttes, outliers of cliffs to the left. Below, they are composed of shales and marls of light blue and slate colors; above, the rocks are buff and gray, and then brown. The buttes are buttressed below, where the azure rocks are seen, and terraced above through the gray and brown beds. A long line of cliffs or rock escarpments separates the table-lands through which Gray Canyon is cut, from the lower plain. The eye can trace these azure beds and cliffs on either side of the river, in a long line extending across its course, until they fade away in the perspective. These cliffs are many miles in length and hundreds of feet high; and all these buttes–great mountain-masses of rock–are dancing and fading away and reappearing, softly moving about,–or so they seem to the eye as seen through the shifting atmosphere.

This afternoon our way is through a valley with cottonwood groves on either side. The river is deep, broad, and quiet. About two hours after noon camp we discover an Indian crossing, where a number of rafts, rudely constructed of logs and bound together by withes, are floating against the bank. On landing, we see evidences that a party of Indians have crossed within a very few days. This is the place where the lamented Gunnison crossed, in the year 1853, when making an exploration for a railroad route to the Pacific coast.

An hour later we run a long rapid and stop at its foot to examine some interesting rocks, deposited by mineral springs that at one time must have existed here, but which are no longer flowing.

July 14.– This morning we pass some curious black bluffs on the right, then two or three short canyons, and then we discover the mouth of the San Rafael, a stream which comes down from the distant mountains in the west. Here we stop for an hour or two and take a short walk up the valley, and find it is a frequent resort for Indians. Arrowheads are scattered about, many of them very beautiful; flint chips are strewn over the ground in great profusion, and the trails are well worn.

Starting after dinner, we pass some beautiful buttes on the left, many of which are very symmetrical. They are chiefly composed of gypsum, of many hues, from light gray to slate color; then pink, purple, and brown beds. Now we enter another canyon. Gradually the walls rise higher and higher as we proceed, and the summit of the canyon is formed of the same beds of orange-colored sandstone. Back from the brink the hollows of the plateau are filled with sands disintegrated from these orange beds. They are of a rich cream color, shading into maroon, everywhere destitute of vegetation, and drifted into long, wave-like ridges.

The course of the river is tortuous, and it nearly doubles upon itself many times. The water is quiet, and constant rowing is necessary to make much headway. Sometimes there is a narrow flood plain between the river and the wall, on one side or the other. Where these long, gentle curves are found, the river washes the very foot of the outer wall. A long peninsula of willow-bordered meadow projects within the curve, and the talus at the foot of the cliff is usually covered with dwarf oaks. The orange-colored sandstone is homogeneous in structure, and the walls are usually vertical, though not very high. Where the river sweeps around a curve under a cliff, a vast hollow dome may be seen, with many caves and deep alcoves, which are greatly admired by the members of the party as we go by.

We camp at night on the left bank.

July 15.--Our camp is in a great bend of the canyon. The curve is to the west and we are on the east side of the river. Just opposite, a little stream comes down through a narrow side canyon. We cross and go up to explore it. At its mouth another lateral canyon enters, in the angle between the former and the main canyon above. Still another enters in the angle between the canyon below and the side canyon first mentioned; so that three side canyons enter at the same point. These canyons are very tortuous, almost closed in from view, and, seen from the opposite side of the river, they appear like three alcoves. We name this Trin-Alcove Bend.

Going up the little stream in the central cove, we pass between high walls of sandstone, and wind about in glens. Springs gush from the rocks at the foot of the walls; narrow passages in the rocks are threaded, caves are entered, and many side canyons are observed.

The right cove is a narrow, winding gorge, with overhanging walls, almost shutting out the light. The left is an amphitheater, turning spirally up, with overhanging shelves. A series of basins filled with water are seen at different altitudes as we pass up; huge rocks are piled below on the right, and overhead there is an arched ceiling. After exploring these alcoves, we recross the river and climb the rounded rocks on the point of the bend. In every direction, as far as we are able to see, naked rocks appear. Buttes are scattered on the landscape, here rounded into cones, there buttressed, columned, and carved in quaint shapes, with deep alcoves and sunken recesses. All about us are basins, excavated in the soft sandstone; and these have been filled by the late rains.

Over the rounded rocks and water pockets we look off on a fine Stretch of river, and beyond are naked rocks and beautiful buttes leading the eye to the Azure Cliffs, and beyond these and above them the Brown Cliffs, and still beyond, mountain peaks; and clouds piled over all.

On we go, after dinner, with quiet water, still compelled to row in order to make fair progress. The canyon is yet very tortuous. About six miles below noon camp we go around a great bend to the right, five miles in length, and come back to a point within a quarter of a mile of where we started. Then we sweep around another great bend to the left, making a circuit of nine miles, and come back to a point within 600 yards of the beginning of the bend. In the two circuits we describe almost the figure 8. The men call it a "bowknot" of river; so we name it Bowknot Bend. The line of the figure is 14 miles in length.

There is an exquisite charm in our ride to-day down this beautiful canyon. It gradually grows deeper with every mile of travel; the walls are symmetrically curved and grandly arched, of a beautiful color, and reflected in the quiet waters in many places so as almost to deceive the eye and suggest to the beholder the thought that he is looking into profound depths. We are all in fine spirits and feel very gay, and the badinage of the men is echoed from wall to wall. Now and then we whistle or shout or discharge a pistol, to listen to the reverberations among the cliffs.

At night we camp on the south side of the great Bowknot, and as we eat supper, which is spread on the beach, we name this Labyrinth Canyon.

July 16.–Still we go down on our winding way. Tower cliffs are passed; then the river widens out for several miles, and meadows are seen on either side between the river and the walls. We name this expansion of the river Tower Park. At two o'clock we emerge from Labyrinth Canyon and go into camp.

July 17.–The line which separates Labyrinth Canyon from the one below is but a line, and at once, this morning, we enter another canyon. The water fills the entire channel, so that nowhere is there room to land. The walls are low, but vertical, and as we proceed they gradually increase in altitude. Running a couple of miles, the river changes its course many degrees toward the east. Just here a little stream comes in on the right and the wall is broken down; so we land and go out to take a view of the surrounding country. We are now down among the buttes, and in a region the surface of which is naked, solid rock–a beautiful red sandstone, forming a smooth, undulating pavement. The Indians call this the Toom'pin Tuweap', or "Rock Land," and sometimes the Toom'pin wunear'l Tuweap', or "Land of Standing Rock."

Off to the south we see a butte in the form of a fallen cross. It is several miles away, but it presents no inconspicuous figure on the landscape and must be many hundreds of feet high, probably more than 2,000. We note its position on our map and name it "The Butte of the Cross."

We continue our journey. In many places the walls, which rise from the water's edge, are overhanging on either side. The stream is still quiet, and we glide along through a strange, weird, grand region. The landscape everywhere, away from the river, is of rock–cliffs of rock, tables of rock, plateaus of rock, terraces of rock, crags of rock–ten thousand strangely carved forms; rocks everywhere, and no vegetation, no soil, no sand. In long, gentle curves the river winds about these rocks.

When thinking of these rocks one must not conceive of piles of boulders or heaps of fragments, but of a whole land of naked rock, with giant forms carved on it: cathedral-shaped buttes, towering hundreds or thousands of feet, cliffs that cannot be scaled, and canyon walls that shrink the river into insignificance, with vast, hollow domes and tall pinnacles and shafts set on the verge overhead; and all highly colored–buff, gray, red, brown, and chocolate–never lichened, never moss-covered, but bare, and often polished.

We pass a place where two bends of the river come together, an intervening rock having been worn away and a new channel formed across. The old channel ran in a great circle around to the right, by what was once a circular peninsula, then an island; then the water left the old channel entirely and passed through the cut, and the old bed of the river is dry. So the great circular rock stands by itself, with precipitous walls all about it, and we find but one place where it can be scaled. Looking from its summit, a long stretch of river is seen, sweeping close to the overhanging cliffs on the right, but having a little meadow between it and the wall on the left. The curve is very gentle and regular. We name this Bonita Bend.

And just here we climb out once more, to take another bearing on The Butte of the Cross. Reaching an eminence from which we can overlook the landscape, we are surprised to find that our butte, with its wonderful form, is indeed two buttes, one so standing in front of the other that from our last point of view it gave the appearance of a cross.

A few miles below Bonita Bend we go out again a mile or two among the rocks, toward the Orange Cliffs, passing over terraces paved with jasper. The cliffs are not far away and we soon reach them, and wander in some deep, painted alcoves which attracted our attention from the river; then we return to our boats.

Late in the afternoon the water becomes swift and our boats make great speed.. An hour of this rapid running brings us to the junction of the Grand and Green, the foot of Stillwater Canyon, as we have named it. These streams-unite in solemn depths, more than 1,200 feet below the general surface of the country. The walls of the lower end of Stillwater Canyon are very beautifully curved, as the river sweeps in its meandering course. The lower end of the canyon through which the Grand comes down is also regular, but much more direct, and we look up this stream and out into the country beyond and obtain glimpses of snow-clad peaks, the summits of a group of mountains known as the Sierra La Sal. Down the Colorado the canyon walls are much broken.

We row around into the Grand and camp on its northwest bank; and here we propose to stay several days, for the purpose of determining the latitude and longitude and the altitude of the walls. Much of the night is spent in making observations with the sextant.

The distance from the mouth of the Uinta to the head of the Canyon of Desolation is 20 3/4 miles. The Canyon of Desolation is 97 miles long; Gray Canyon, 36 miles. The course of the river through Gunnison Valley is 27 1/4 miles; Labyrinth Canyon, 62 1/2 miles.

In the Canyon of Desolation the highest rocks immediately over the river are about 2,400 feet. This is at Log Cabin Cliff. The highest part of the terrace is near the brink of the Brown Cliffs. Climbing the immediate walls of the canyon and passing back to the canyon terrace and climbing that, we find the altitude above the river to be 3,300 feet. The lower end of Gray Canyon is about 2,000 feet; the lower end of Labyrinth Canyon, 1,300 feet.

Stillwater Canyon is 42 3/4 miles long; the highest walls, 1,300 feet.

CHAPTER X. FROM THE JUNCTION OF THE GRAND AND GREEN TO THE MOUTH OF THE LITTLE COLORADO.

July 18.–The day is spent in obtaining the time and spreading our rations, which we find are badly injured. The flour has been wet and dried so many times that it is all musty and full of hard lumps. We make a sieve of mosquito netting and run our flour through, it, losing more than 200 pounds by the process. Our losses, by the wrecking of the "No Name," and by various mishaps since, together with the amount thrown away to-day, leave us little more than two months' supplies, and to make them last thus long we must be fortunate enough to lose no more.

We drag our boats on shore and turn them over to recalk and pitch them, and Sumner is engaged in repairing barometers. While we are here for a day or two, resting, we propose to put everything in the best shape for a vigorous campaign.

July 19.–Bradley and I start this morning to climb the left wall below the junction. The way we have selected is up a gulch. Climbing for an hour over and among the rocks, we find ourselves in a vast amphitheater and our way cut off. We clamber around to the left for half an hour, until we find that we cannot go up in that direction. Then we try the rocks around to the right and discover a narrow shelf nearly half a mile long. In some places this is so wide that we pass along with ease; in others it is so narrow and sloping that we are compelled to lie down and crawl. We can look over the edge of the shelf, down 800 feet, and see the river rolling and plunging among the rocks. Looking up 500 feet to the brink of the cliff, it seems to blend with the sky. We continue along until we come to a point where the wall is again broken down. Up we climb. On the right there is a narrow, mural point of rocks, extending toward the river, 200 or 300 feet high and 600 or 800 feet long. We come back to where this sets in and find it cut off from the main wall by a great crevice. Into this we pass; and now a long, narrow rock is between us and the river. The rock itself is split longitudinally and transversely; and the rains on the surface above have run down through the crevices and gathered into channels below and then run off into the river. The crevices are usually narrow above and, by erosion of the streams, wider below, forming a network of "caves", each cave having a narrow, winding skylight up through the rocks. We wander among these corridors for an hour or two, but find no place where the rocks are broken down so that we can climb up. At last we determine to attempt a passage by a crevice, and select one which we think is wide enough to admit of the passage of our bodies and yet narrow enough to climb out by pressing our hands and feet against the walls. So we climb as men would out of a well. Bradley climbs first; I hand him the barometer, then climb over his head and he hands me the barometer. So we pass each other alternately until we emerge from the fissure, out on the summit of the rock. And what a world of grandeur is spread before us! Below is the canyon through which the Colorado runs. We can trace its course for miles, and at points catch glimpses of the river. From the northwest comes the Green in a narrow winding gorge. From the northeast comes the Grand, through a canyon that seems bottomless from where we stand. Away to the west are lines of cliffs and ledges of rock–not such ledges as the reader may have seen where the quarryman splits his blocks, but ledges from which the gods might quarry mountains that, rolled out on the plain below, would stand a lofty range; and not such cliffs as the reader may have seen where the swallow builds its nest, but cliffs where the soaring eagle is lost to view ere he reaches the summit. Between us and the distant cliffs are the strangely carved and pinnacled rocks of the Toom'pin wunear' Tuweap'. On the summit of the opposite wall of the canyon are rock forms that we do not understand. Away to the east a group of eruptive mountains are seen–the Sierra La Sal, which we first saw two days ago through the canyon of the Grand. Their slopes are covered with pines, and deep gulches are flanked with great crags, and snow fields are seen near the summits. So the mountains are in uniform,–green, gray, and silver. Wherever we look there is but a wilderness of rocks,–deep gorges where the rivers are lost below cliffs and towers and pinnacles, and ten thousand strangely carved forms in every direction, and beyond them mountains blending with the clouds.

Now we return to camp. While eating supper we very naturally speak of better fare, as musty bread and spoiled bacon are not palatable. Soon I see Hawkins down by the boat, taking up the sextant–rather a strange proceeding for him–and I question him concerning it. He replies that he is trying to find the latitude and longitude of the nearest pie.

July 20.–This morning Captain Powell and I go out to climb the west wall of the canyon, for the purpose of examining the strange rocks seen yesterday from the other side. Two hours bring us to the top, at a point between the Green and Colorado overlooking the junction of the rivers.

A long neck of rock extends toward the mouth of the Grand. Out on this we walk, crossing a great number of deep crevices. Usually the smooth rock slopes down to the fissure on either side. Sometimes it is an interesting question to us whether the slope is not so steep that we cannot stand on it. Sometimes, starting down, we are compelled to go on, and when we measure the crevice with our eye from above we are not always sure that it is not too wide for a jump. Probably the slopes would not be difficult if there was not a fissure at the lower end; nor would the fissures cause fear if they were but a few feet deep. It is curious how a little obstacle becomes a great obstruction when a misstep would land a man in the bottom of a deep chasm. Climbing the face of a cliff, a man will without hesitancy walk along a step or shelf but a few inches wide if the landing is but ten feet below, but if the foot of the cliff is a thousand feet down he will prefer to crawl along the shelf. At last our way is cut off by a fissure so deep and wide that we cannot pass it. Then we turn and walk back into the country, over the smooth, naked sandstone, without vegetation, except that here and there dwarf cedars and piñón pines have found a footing in the huge cracks. There are great basins in the rock, holding water,–some but a few gallons, others hundreds of barrels.

The day is spent in walking about through these strange scenes. A narrow gulch is cut into the wall of the main canyon. Follow this up and the climb is rapid, as if going up a mountain side, for the gulch heads but a few hundred or a few thousand yards from the wall. But this gulch has its side gulches, and as the summit is approached a group of radiating canyons is found. The spaces drained by these little canyons are terraced, and are, to a greater or less extent, of the form of amphitheaters, though some are oblong and some rather irregular. Usually the spaces drained by any two of these little side canyons are separated by a narrow wall, 100, 200, or 300 feet high, and often but a few feet in thickness. Sometimes the wall is broken into a line of pyramids above and still remains a wall below. There are a number of these gulches which break the wall of the main canyon of the Green, each one having its system of side canyons and amphitheaters, inclosed by walls or lines of pinnacles. The course of the Green at this point is approximately at right angles to that of the Colorado, and on the brink of the latter canyon we find the same system of terraced and walled glens. The walls and pinnacles and towers are of sandstone, homogeneous in structure but not in color, as they show broad bands of red, buff, and gray. This painting of the rocks, dividing them into sections, increases their apparent height. In some places these terraced and walled glens along the Colorado have coalesced with those along the Green; that is, the intervening walls are broken down. It is very rarely that a loose rock is seen. The sand is washed off, so that the walls, terraces, and slopes of the glens are all of smooth sandstone.

In the walls themselves curious caves and channels have been carved. In some places there are little stairways up the walls; in others, the walls present what are known as royal arches; and so we wander through glens and among pinnacles and climb the walls from early morn until late in the afternoon.

July 21.– We start this morning on the Colorado. The river is rough, and bad rapids in close succession are found. Two very hard portages are made during the forenoon. After dinner, in running a rapid, the "Emma Dean" is swamped and we are thrown into the river; we cling to the boat, and in the first quiet water below she is righted and bailed out; but three oars are lost in this mishap. The larger boats land above the dangerous place, and we make a portage, which occupies all the afternoon. We camp at night on the rocks on the left bank, and can scarcely find room to lie down.

July 22.–This morning we continue our journey, though short of oars. There is no timber growing on the walls within our reach and no driftwood along the banks, so we are compelled to go on until something suitable can be found. A mile and three quarters below, we find a huge pile of driftwood, among which are some cottonwood logs. From these we select one which we think the best, and the men are set at work sawing oars. Our boats are leaking again, from the strains received in the bad rapids yesterday, so after dinner they are turned over and some of the men calk them.

Captain Powell and I go out to climb the wall to the east, for we can see dwarf pines above, and it is our purpose to collect the resin which oozes from them, to use in pitching our boats. We take a barometer with us and find that the walls are becoming higher, for now they register an altitude above the river of nearly 1,500 feet.

July 23.–On starting, we come at once to difficult rapids and falls, that in many places are more abrupt than in any of the canyons through which we have passed, and we decide to name this Cataract Canyon. From morning until noon the course of the river is to the west; the scenery is grand, with rapids, and falls below, and walls above, beset with crags and pinnacles. Just at noon we wheel again to the south and go into camp for dinner.

While the cook is preparing it, Bradley, Captain Powell, and I go up into a side canyon that comes in at this point. We enter through a very narrow passage, having to wade along the course of a little stream until a cascade interrupts our progress. Then we climb to the right for a hundred feet until we reach a little shelf, along which we pass, walking with great care, for it is narrow; thus we pass around the fall. Here the gorge widens into a spacious, sky-roofed chamber. In the farther end is a beautiful grove of cottonwoods, and between us and the cotton-woods the little stream widens out into three clear lakelets with bottoms of smooth rock. Beyond the cottonwoods the brook tumbles in a series of white, shining cascades from heights that seem immeasurable. Turning around, we can look through the cleft through which we came and see the river with towering walls beyond. What a chamber for a resting-place is this! hewn from the solid rock, the heavens for a ceiling, cascade fountains within, a grove in the conservatory, clear lakelets for a refreshing bath, and an outlook through the doorway on a raging river, with cliffs and mountains beyond.

Our way after dinner is through a gorge, grand beyond description. The walls are nearly vertical, the river broad and swift, but free from rocks and falls. From the edge of the water to the brink of the cliffs it is 1,600 to 1,800 feet. At this great depth the river rolls in solemn majesty. The cliffs are reflected from the more quiet river, and we seem to be in the depths of the earth, and yet we can look down into waters that reflect a bottomless abyss. Early in the afternoon we arrive at the head of more rapids and falls, but, wearied with past work, we determine to rest, so go into camp, and the afternoon and evening are spent by the men in discussing the probabilities of successfully navigating the river below. The barometric records are examined to see what descent we have made since we left the mouth of the Grand, and what descent since we left the Pacific Railroad, and what fall there yet must be to the river ere we reach the end of the great canyons. The conclusion at which the men arrive seems to be about this: that there are great descents yet to be made, but if they are distributed in rapids and short falls, as they have been heretofore, we shall be able to overcome them; but may be we shall come to a fall in these canyons which we cannot pass, where the walls rise from the water's edge, so that we cannot land, and where the water is so swift that we cannot return. Such places have been found, except that the falls were not so great but that we could run them with safety. How will it be in the future t So they speculate over the serious probabilities in jesting mood.

July 24.–We examine the rapids below. Large rocks have fallen from the walls–great, angular blocks, which have rolled down the talus and are strewn along the channel. We are compelled to make three portages in succession, the distance being less than three fourths of a mile, with a fall of 75 feet. Among these rocks, in chutes, whirlpools, and great waves, with rushing breakers and foam, the water finds its way, still tumbling down. We stop for the night only three fourths of a mile below the last camp. A very hard day's work has been done, and at evening I sit on a rock by the edge of the river and look at the water and listen to its roar. Hours ago deep shadows settled into the canyon, as the sun passed behind the cliffs. Now, doubtless, the sun has gone down, for we can see no glint of light on the crags above. Darkness is coming on; but the waves are rolling with crests of foam so white they seem almost to give a light of their own. Near by, a chute of water strikes the foot of a great block of limestone 50 feet high, and the waters pile up against it and roll back. Where there are sunken rocks the water heaps up in mounds, or even in cones. At a point where rocks come very near the surface, the water forms a chute above, strikes, and is shot up 10 or 15 feet, and piles back in gentle curves, as in a fountain; and on the river tumbles and rolls.

July 25.–Still more rapids and falls to-day. In one, the "Emma Dean" is caught in a whirlpool and set spinning about, and it is with great difficulty we are able to get out of it with only the loss of an oar. At noon another is made; and on we go, running some of the rapids, letting down with lines past others, and making two short portages. We camp on the right bank, hungry and tired.

July 26.–We run a short distance this morning and go into camp to make oars and repair boats and barometers. The walls of the canyon have been steadily increasing in altitude to this point, and now they are more than 2,000 feet high. In many places they are vertical from the water's edge; in others there is a talus between the river and the foot of the cliff; and they are often broken down by side canyons. It is probable that the river is nearly as low now as it is ever found. High-water mark can be observed 40, 50, 60, or 100 feet above its present stage. Sometimes logs and driftwood are seen wedged into the crevices over-head, where floods have carried them.

About ten o'clock, Powell, Bradley, Howland, Hall, and I start up a side canyon to the east. We soon come to pools of water; then to a brook, which is lost in the sands below; and passing up the brook, we see that the canyon narrows, the walls close in and are often overhanging, and at last we find ourselves in a vast amphitheater, with a pool of deep, clear, cold water on the bottom. At first our way seems cut off; but we soon discover a little shelf, along which we climb, and, passing beyond the pool, walk a hundred yards or more, turn to the right, and find ourselves in another dome-shaped amphitheater. There is a winding cleft at the top, reaching out to the country above, nearly 2,000 feet overhead. The rounded, basin-shaped bottom is filled with water to the foot of the walls. There is no shelf by which we can pass around the foot. If we swim across we meet with a face of rock hundreds of feet high, over which a little rill glides, and it will be impossible to climb. So we can go no farther up this canyon. Then we turn back and examine the walls on either side carefully, to discover, if possible, some way of climbing out. In this search every man takes his own course, and we are scattered. I almost abandon the idea of getting out and am engaged in searching for fossils, when I discover, on the north, a broken place lip which it may be possible to climb. The way for a distance is up a slide of rocks; then up an irregular amphitheater, on points that form steps and give handhold; and then I reach a little shelf, along which I walk, and discover a vertical fissure parallel to the face of the wall and reaching to a higher shelf. This fissure is narrow and I try to climb up to the bench, which is about 40 feet overhead. I have a barometer on my back, which rather impedes my climbing. The walls of the fissure are of smooth limestone, offering neither foothold nor handhold. So I support myself by pressing my back against one wall and my knees against the other, and in this way lift my body, in a shuffling manner, a few inches at a time, until I have made perhaps 25 feet of the distance, when the crevice widens a little and I cannot press my knees against the rock in front with sufficient power to give me support in lifting my body; so I try to go back. This I cannot do without falling. So I struggle along sidewise farther into the crevice, where it narrows. But by this time my muscles are exhausted, and I cannot climb longer; so I move still a little farther into the crevice, where it is so narrow and wedging that I can lie in it, and there I rest. Five or ten minutes of this relief, and up once more I go, and reach the bench above. On this I can walk for a quarter of a mile, till I come to a place where the wall is again broken down, so I can climb up still farther; and in an hour I reach the summit. I hang up my barometer to give it a few minutes' time to settle, and occupy myself in collecting resin from the pinon pines, which are found in great abundance. One of the principal objects in making this climb was to get this resin for the purpose of smearing our boats; but I have with me no means of carrying it down. The day is very hot and my coat was left in camp, so I have no linings to tear out. Then it occurs to me to cut off the sleeve of my shirt and tie it up at one end, and in this little sack I collect about a gallon of pitch. After taking observations for altitude, I wander back on the rock for an hour or two, when suddenly I notice that a storm is coming from the south. I seek a shelter in the rocks; but when the storm bursts, it comes down as a flood from the heavens,–not with gentle drops at first, slowly increasing in quantity, but as if suddenly poured out. I am thoroughly drenched and almost washed away. It lasts not more than half an hour, when the clouds sweep by to the north and I have sunshine again.

In the meantime I have discovered a better way of getting down, and start for camp, making the greatest haste possible. On reaching the bottom of the side canyon, I find a thousand streams rolling down the cliffs on every side, carrying with them red sand; and these all unite in the canyon below in one great stream of red mud.

Traveling as fast as I can run, I soon reach the foot of the stream, for the rain did not reach the lower end of the canyon and the water is running down a dry bed of sand; and although it conies in waves several feet high and 15 or 20 feet in width, the sands soak it up and it is lost. But wave follows wave and rolls along and is swallowed up; and still the floods come on from above. I find that I can travel faster than the stream; so I hasten to camp and tell the men there is a river coming down the canyon. We carry our camp equipage hastily from the bank to where we think it will be above the water. Then we stand by and see the river roll on to join the Colorado. Great quantities of gypsum are found at the bottom of the gorge; so we name it Gypsum Canyon.

July 27.–We have more rapids and falls until noon; then we come to a narrow place in the canyon, with vertical walls for several hundred feet, above which are steep steps and sloping rocks back to the summits. The river is very narrow, and we make our way with great care and much anxiety, hugging the wall on the left and carefully examining the way before us.

Late in the afternoon we pass to the left around a sharp point, which is somewhat broken down near the foot, and discover a flock of mountain sheep on the rocks more than a hundred feet above us. We land quickly in a cove out of sight, and away go all the hunters with their guns, for the sheep have not discovered us. Soon we hear firing, and those of us who have remained in the boats climb up to see what success the hunters have had. One sheep has been killed, and two of the men are still pursuing them. In a few minutes we hear firing again, and the next moment down come the flock clattering over the rocks within 20 yards of us. One of the hunters seizes his gun and brings a second sheep down, and the next minute the remainder of the flock is lost behind the rocks. We all give chase; but it is impossible to follow their tracks over the naked rock, and we see them no more. Where they went out of this rock-walled canyon is a mystery, for we can see no way of escape. Doubtless, if we could spare the time for the search, we should find a gulch up which they ran.

We lash our prizes to the deck of one of the boats and go on for a short distance; but fresh meat is too tempting for us, and we stop early to have a feast. And a feast it is! Two fine young sheep! We care not for bread or beans or dried apples to-night; coffee and mutton are all we ask.

July 28.–We make two portages this morning, one of them very long. During the afternoon we run a chute more than half a mile in length, narrow and rapid. This chute has a floor of marble; the rocks dip in the direction in which we are going, and the fall of the stream conforms to the inclination of the beds; so we float on water that is gliding down an inclined plane. At the foot of the chute the river turns sharply to the right and the water rolls up against a rock which from above seems to stand directly athwart its course. As we approach it we pull with all our power to the right, but it seems impossible to avoid being carried headlong against the cliff; we are carried up high on the waves–but not against the rock, for the rebounding water strikes us and we are beaten back and pass on with safety, except that we get a good drenching.

After this the walls suddenly close in, so that the canyon is narrower than we have ever known it. The water fills it from wall to wall, giving us no landing-place at the foot of the cliff; the river is very swift and the canyon very tortuous, so that we can see but a few hundred yards ahead; the walls tower over us, often overhanging so as almost to shut out the light. I stand on deck, watching with intense anxiety, lest this may lead us into some danger; but we glide along, with no obstruction, no falls, no rocks, and in a mile and a half emerge from the narrow gorge into a more open and broken portion of the canyon. Now that it is past, it seems a very simple thing indeed to run through such a place, but the fear of what might be ahead made a deep impression on us.

At three o'clock we arrive at the foot of Cataract Canyon. Here a long canyon valley comes down from the east, and the river turns sharply to the west in a continuation of the line of the lateral valley. In the bend on the right vast numbers of crags and pinnacles and tower-shaped rocks are seen. We call it Mille Crag Bend.

And now we wheel into another canyon, on swift water unobstructed by rocks. This new canyon is very narrow and very straight, with walls vertical below and terraced above. Where we enter it the brink of the cliff is 1,300 feet above the water, but the rocks dip to the west, and as the course of the canyon is in that direction the walls are seen slowly to decrease in altitude. Floating down this narrow channel and looking out through the canyon crevice away in the distance, the river is seen to turn again to the left, and beyond this point, away many miles, a great mountain is seen. Still floating down, we see other mountains, now on the right, now on the left, until a great mountain range is unfolded to view. We name this Narrow Canyon, and it terminates at the bend of the river below.

As we go down to this point we discover the mouth of a stream which enters from the right. Into this our little boat is turned. The water is exceedingly muddy and has an unpleasant odor. One of the men in the boat following, seeing what we have done, shouts to 'Dunn and asks whether it is a trout stream. Dunn replies, much disgusted, that it is "a dirty devil," and by this name the river is to be known hereafter.

Some of us go out for half a mile and climb a butte to the north.

The course of the Dirty Devil River can be traced for many miles. It comes down through a very narrow canyon, and beyond it, to the southwest, there is a long line of cliffs, with a broad terrace, or bench, between it and the brink of the canyon, and beyond these cliffs is situated the range of mountains seen as we came down Narrow Canyon. Looking up the Colorado, the chasm through which it runs can be seen, but we cannot see down to its waters. The whole country is a region of naked rock of many colors, with cliffs and buttes about us and towering mountains in the distance.

July 29.–We enter a canyon to-day, with low, red walls. A short distance below its head we discover the ruins of an old building on the left wall. There is a narrow plain between the river and the wall just here, and on the brink of a rock 200 feet high stands this old house. Its walls are of stone, laid in mortar with much regularity. It was probably built three stories high; the lower story is yet almost intact; the second is much broken down, and scarcely anything is left of the third. Great quantities of flint chips are found on the rocks near by, and many arrowheads, some perfect, others broken; and fragments of pottery are strewn about in great profusion. On the face of the cliff, under the building and along down the river for 200 or 300 yards, there are many etchings. Two hours are given to the examination of these interesting ruins; then we run down fifteen miles farther, and discover another group. The principal building was situated on the summit of the hill.

A part of the walls are standing, to the height of eight or ten feet, and the mortar yet remains in some places. The house was in the shape of an L, with five rooms on the ground floor,–one in the angle and two in each extension. In the space in the angle there is a deep excavation. From what we know of the people in the Province of Tusayan, who are, doubtless, of the same race as the former inhabitants of these ruins, we conclude that this was a kiva, or underground chamber in which their religious ceremonies were performed.

We leave these ruins and run down two or three miles and go into camp about mid-afternoon. And now I climb the wall and go out into the back country for a walk.

The sandstone through which the canyon is cut is red and homogeneous, being the same as that through which Labyrinth Canyon runs. The smooth, naked rock stretches out on either side of the river for many miles, but curiously carved mounds and cones are scattered everywhere and deep holes are worn out. Many of these pockets are filled with water. In one of these holes or wells, 20 feet deep, I find a tree growing. The excavation is so narrow that I can step from its brink to a limb on the tree and descend to the bottom of the well down a growing ladder. Many of these pockets are potholes, being found in the courses of little rills or brooks that run during the rains which occasionally fall in this region; and often a few harder rocks, which evidently assisted in their excavation, can be found in their bottoms. Others, which are shallower, are not so easily explained. Perhaps where they are found softer spots existed in the sandstone, places that yielded more readily to atmospheric degradation, the loose sands being carried away by the winds.

Just before sundown I attempt to climb a rounded eminence, from which I hope to obtain a good outlook on the surrounding country. It is formed of smooth mounds, piled one above another. Up these I climb, winding here and there to find a practicable way, until near the summit they become too steep for me to proceed. I search about a few minutes for an easier way, when I am surprised at finding a stairway, evidently cut in the rock by hands. At one place, where there is a vertical wall of 10 or 12 feet, I find an old, rickety ladder. It may be that this was a watchtower of that ancient people whose homes we have found in ruins. On many of the tributaries of the Colorado, I have heretofore examined their deserted dwellings. Those that show evidences of being built during the latter part of their occupation of the country are usually placed on the most inaccessible cliffs. Sometimes the mouths of caves have been walled across, and there are many other evidences to show their anxiety to secure defensible positions. Probably the nomadic tribes were sweeping down upon them and they resorted to these cliffs and canyons for safety. It is not unreasonable to suppose that this orange mound was used as a watchtower. Here I stand, where these now lost people stood centuries ago, and look over this strange country, gazing off to great mountains in the northwest which are slowly disappearing under cover of the night; and then I return to camp. It is no easy task to find my way down the wall in the darkness, and I clamber about until it is nearly midnight when camp is reached.

July 30.–We make good progress to-day, as the water, though smooth, is swift. Sometimes the canyon walls are vertical to the top; sometimes they are vertical below and have a mound-covered slope above; in other places the slope, with its mounds, comes down to the water's edge.

Still proceeding on our way, we find that the orange sandstone is cut in two by a group of firm, calcareous strata, and the lower bed is underlaid by soft, gypsiferous shales. Sometimes the upper homogeneous bed is a smooth, vertical wall, but usually it is carved with mounds, with gently meandering valley lines. The lower bed, yielding to gravity, as the softer shales below work, out into the river, breaks into angular surfaces, often having a columnar appearance. One could almost imagine that the walls had been carved with a purpose, to represent giant architectural forms. In the deep recesses of the walls we find springs, with mosses and ferns on the moistened sandstone.

July 31.–We have a cool, pleasant ride to-day through this part of the canyon. The walls are steadily increasing in altitude, the curves are gentle, and often the river sweeps by an arc of vertical wall, smooth and unbroken, and then by a curve that is variegated by royal arches, mossy alcoves, deep, beautiful glens, and painted grottoes. Soon after dinner we discover the mouth of the San Juan, where we camp. The remainder of the afternoon is given to hunting some way by which we can climb out of the canyon; but it ends in failure.

August 1.–We drop down two miles this morning and go into camp again. There is a low, willow-covered strip of land along the walls on the east. Across this we walk, to explore an alcove which we see from the river. On entering, we find a little grove of box-elder and cotton-wood trees, and turning to the right, we find ourselves in a vast chamber, carved out of the rock. At the upper end there is a clear, deep pool of water, bordered with verdure. Standing by the side of this, we can see the grove at the entrance. The chamber is more than 200 feet high, 500 feet long, and 200 feet wide. Through the ceiling, and on through the rocks for a thousand feet above, there is a narrow, winding skylight; and this is all carved out by a little stream which runs only during the few showers that fall now and then in this arid country. The waters from the bare rocks back of the canyon, gathering rapidly into a small channel, have eroded a deep side canyon, through which they run until they fall into the farther end of this chamber. The rock at the ceiling is hard, the rock below, very soft and friable; and having cut through the upper and harder portion down into the lower and softer, the stream has washed out these friable sandstones; and thus the chamber has been excavated.

Here we bring our camp. When "Old Shady" sings us a song at night, we are pleased to find that this hollow in the rock is filled with sweet sounds. It was doubtless made for an academy of music by its storm-born architect; so we name it Music Temple.

August 2.–We still keep our camp in Music Temple to-day. I wish to obtain a view of the adjacent country, if possible; so, early in the morning the men take me across the river, and I pass along by the foot of the cliff half a mile up stream and then climb, first up broken ledges, then 200 or 300 yards up a smooth, sloping rock, and then pass out on a narrow ridge. Still, I find I have not attained an altitude from which I can overlook the region outside of the canyon; and so I descend into a little gulch and climb again to a higher ridge, all the way along naked sandstone, and at last I reach a point of commanding view. I can look several miles up the San Juan, and a long distance up the Colorado; and away to the northwest I can see the Henry Mountains; to the northeast, the Sierra La Sal; to the southeast, unknown mountains; and to the southwest, the meandering of the canyon. Then I return to the bank of the river. We sleep again in Music Temple.

August 3.–Start early this morning. The features of this canyon are greatly diversified. Still vertical walls at times. These are usually found to stand above great curves. The river, sweeping around these bends, undermines the cliffs in places. Sometimes the rocks are overhanging; in other curves, curious, narrow glens are found. Through these we climb, by a rough stairway, perhaps several hundred feet, to where a spring bursts out from under an overhanging cliff, and where cottonwoods and willows stand, while along the curves of the brooklet oaks grow, and other rich vegetation is seen, in marked contrast to the general appearance of naked rock. We call these Oak Glens.

Other wonderful features are the many side canyons or gorges that we pass. Sometimes we stop to explore these for a short distance. In some places their walls are much nearer each other above than below, so that they look somewhat like caves or chambers in the rocks. Usually, in going up such a gorge, we find beautiful vegetation; but our way is often cut off by deep basins, or "potholes," as they are called.

On the walls, and back many miles into the country, numbers of monument-shaped buttes are observed. So we have a curious ensemble of wonderful features–carved walls, royal arches, glens, alcove gulches, mounds, and monuments. From which of these features shall we select a name? We decide to call it Glen Canyon.

Past these towering monuments, past these mounded billows of orange sandstone, past these oak-set glens, past these fern-decked alcoves, past these mural curves, we glide hour after hour, stopping now and then, as our attention is arrested by some new wonder, until we reach a point which is historic.

In the year 1776, Father Escalante, a Spanish priest, made an expedition from Santa Fe to the northwest, crossing the Grand and Green, and then passing down along the Wasatch Mountains and the southern plateaus until he reached the Rio Virgen. His intention was to cross to the Mission of Monterey; but, from information received from the Indians, he decided that the route was impracticable. Not wishing to return to Santa Fe over the circuitous route by which he had just traveled, he attempted to go by one more direct, which led him across the Colorado at a point known as El Vado de los Padres. From the description which we have read, we are enabled to determine the place. A little stream comes down through a very narrow side canyon from the west. It was down this that he came, and our boats are lying at the point where the ford crosses. A well-beaten Indian trail is seen here yet. Between the cliff and the river there is a little meadow. The ashes of many camp fires are seen, and the bones of numbers of cattle are bleaching on the grass. For several years the Navajos have raided on the Mormons that dwell in the valleys to the west, and they doubtless cross frequently at this ford with their stolen cattle.

August 4.–To-day the walls grow higher and the canyon much narrower. Monuments are still seen on either side; beautiful glens and alcoves and gorges and side canyons are yet found. After dinner we find the river making a sudden turn to the northwest and the whole character of the canyon changed. The walls are many hundreds of feet higher, and the rocks are chiefly variegated shales of beautiful colors–creamy orange above, then bright vermilion, and below, purple and chocolate beds, with green and yellow sands. We run four miles through this, in a direction a little to the west of north, wheel again to the west, and pass into a portion of the canyon where the characteristics are more like those above the bend. At night we stop at the mouth of a creek coming in from the right, and suppose it to be the Paria, which was described to me last year by a Mormon missionary. Here the canyon terminates abruptly in a line of cliffs, which stretches from either side across the river.

August 5.–With some feeling of anxiety we enter a new canyon this morning. We have learned to observe closely the texture of the rock. In softer strata we have a quiet river, in harder we find rapids and falls. Below us are the limestones and hard sandstones which we found in Cataract Canyon. This bodes toil and danger. Besides the texture of the rocks, there is another condition which affects the character of the channel, as we have found by experience. Where the strata are horizontal the river is often quiet, and, even though it may be very swift in places, no great obstacles are found. Where the rocks incline in the direction traveled, the river usually sweeps with great velocity, but still has few rapids and falls. But where the rocks dip up stream and the river cuts obliquely across the upturned formations, harder strata above and softer below, we have rapids and falls. Into hard rocks and into rocks dipping up stream we pass this morning and start on a long, rocky, mad rapid. On the left there is a vertical rock, and down by this cliff and around to the left we glide, tossed just enough by the waves to appreciate the rate at which we are traveling.

The canyon is narrow, with vertical walls, which gradually grow higher. More rapids and falls are found. We come to one with a drop of sixteen feet, around which we make a portage, and then stop for dinner. Then a run of two miles, and another portage, long and difficult; then we camp for the night on a bank of sand.

August 6.–Canyon walls, still higher and higher, as we go down through strata. There is a steep talus at the foot of the cliff, and in some places the upper parts of the walls are terraced.

About ten o'clock we come to a place where the river occupies the entire channel and the walls are vertical from the water's edge. We see a fall below and row up against the cliff. There is a little shelf, or rather a horizontal crevice, a few feet over our heads. One man stands on the deck of the boat, another climbs on his shoulders, and then into the crevice. Then we pass him a line, and two or three others, with myself, follow; then we pass along the crevice until it becomes a shelf, as the upper part, or roof, is broken off. On this we walk for a short distance, slowly climbing all the way, until we reach a point where the shelf is broken off, and we can pass no farther. So we go back to the boat, cross the stream, and get some logs that have lodged in the rocks, bring them to our side, pass them along the crevice and shelf, and bridge over the broken place. Then we go on to a point over the falls, but do not obtain a satisfactory view. So we climb out to the top of the wall and walk along to find a point below the fall from which it can be seen. From this point it seems possible to let down our boats with lines to the head of the rapids, and then make a portage; so we return, row down by the side of the cliff as far as we dare, and fasten one of the boats to a rock. Then we let down another boat to the end of its line beyond the first, and the third boat to the end of its line below the second, which brings it to the head of the fall and under an overhanging rock. Then the upper boat, in obedience to a signal, lets go; we pull in the line and catch the nearest boat as it comes, and then the last. The portage follows.

We go into camp early this afternoon at a place where it seems possible to climb out, and the evening is spent in "making observations for time."

August 7.–The almanac tells us that we are to have an eclipse of the sun to-day; so Captain Powell and myself start early, taking our instruments with us for the purpose of making observations on the eclipse to determine our longitude. Arriving at the summit, after four hours' hard climbing to attain 2,300 feet in height, we hurriedly build a platform of rocks on which to place our instruments, and quietly wait for the eclipse; but clouds come on and rain falls, and sun and moon are obscured.

Much disappointed, we start on our return to camp, but it is late and the clouds make the night very dark. We feel our way down among the rocks with great care for two or three hours, making slow progress indeed. At last we lose our way and dare proceed no farther. The rain comes down in torrents and we can find no shelter. We can neither climb up nor go down, and in the darkness dare not move about; so we sit and "weather out" the night.

August 8.–Daylight comes after a long, oh, how long! a night, and we soon reach camp. After breakfast we start again, and make two portages during the forenoon.

The limestone of this canyon is often polished, and makes a beautiful marble. Sometimes the rocks are of many colors–white, gray, pink, and purple, with saffron tints. It is with very great labor that we make progress, meeting with many obstructions, running rapids, letting down our boats with lines from rock to rock, and sometimes carrying boats and cargoes around bad places. We camp at night, just after a hard portage, under an overhanging wall, glad to find shelter from the rain. We have to search for some time to find a few sticks of driftwood, just sufficient to boil a cup of coffee.

The water sweeps rapidly in this elbow of river, and has cut its way under the rock, excavating a vast half-circular chamber, which, if utilized for a theater, would give sitting to 50,000 people. Objection might be raised against it, however, for at high water the floor is covered with a raging flood.

August 9.–And now the scenery is on a grand scale. The walls of the canyon, 2,500 feet high, are of marble, of many beautiful colors, often polished below by the waves, and sometimes far up the sides, where showers have washed the sands over the cliffs. At one place I have a walk for more than a mile on a marble pavement, all polished and fretted with strange devices and embossed in a thousand fantastic patterns. Through a cleft in the wall the sun shines on this pavement and it gleams in iridescent beauty.

I pass up into the cleft. It is very narrow, with a succession of pools standing at higher levels as I go back. The water in these pools is clear and cool, coming down from springs. Then I return to the pavement, which is but a terrace or bench, over which the river runs at its flood, but left bare at present. Along the pavement in many places are basins of clear water, in strange contrast to the red mud of the river. At length I come to the end of this marble terrace and take again to the boat.

Riding down a short distance, a beautiful view is presented. The river turns sharply to the east and seems inclosed by a wall set with a million brilliant gems. What can it mean? Every eye is engaged, every one wonders. On coming nearer we find fountains bursting from the rock high overhead, and the spray in the sunshine forms the gems which bedeck the wall. The rocks below the fountain are covered with mosses and ferns and many beautiful flowering plants. We name it Vasey's Paradise, in honor of the botanist who traveled with us last year.

We pass many side canyons to-day that are dark, gloomy passages back into the heart of the rocks that form the plateau through which this canyon is cut. It rains again this afternoon. Scarcely do the first drops fall when little rills run down the walls. As the storm comes on, the little rills increase in size, until great streams are formed. Although the walls of the canyon are chiefly limestone, the adjacent country is of red sandstone; and now the waters, loaded with these sands, come down in rivers of bright red mud, leaping over the walls in innumerable cascades. It is plain now how these walls are polished in many places.

At last the storm ceases and we go on. We have cut through the sandstones and limestones met in the upper part of the canyon, and through one great bed of marble a thousand feet in thickness. In this, great numbers of caves are hollowed out, and carvings are seen which suggest architectural forms, though on a scale so grand that architectural terms belittle them. As this great bed forms a distinctive feature of the canyon, we call it Marble Canyon.

It is a peculiar feature of these walls that many projections are set out into the river, as if the wall was buttressed for support. The walls themselves are half a mile high, and these buttresses are on a corresponding scale, jutting into the river scores of feet. In the recesses between these projections there are quiet bays, except at the foot of a rapid, when there are dancing eddies or whirlpools. Sometimes these alcoves have caves at the back, giving them the appearance of great depth. Then other caves are seen above, forming vast dome-shaped chambers. The walls and buttresses and chambers are all of marble.

The river is now quiet; the canyon wider. Above, when the river is at its flood, the waters gorge up, so that the difference between high and low water mark is often 50 or even 70 feet, but here high-water mark is not more than 20 feet above the present stage of the river. Sometimes there is a narrow flood plain between the water and the wall. Here we first discover mesquite shrubs,–small trees with finely divided leaves and pods, somewhat like the locust.

August 10.–Walls still higher; water swift again. We pass several broad, ragged canyons on our right, and up through these we catch glimpses of a forest-clad plateau, miles away to the west.

At two o'clock we reach the mouth of the Colorado Chiquito. This stream enters through a canyon on a scale quite as grand as that of the Colorado itself. It is a very small river and exceedingly muddy and saline. I walk up the stream three or four miles this afternoon, crossing and recrossing where I can easily wade it. Then I climb several hundred feet at one place, and can see for several miles up the chasm through which the river runs. On my way back I kill two rattlesnakes, and find on my arrival that another has been killed just at camp.

August 11.–We remain at this point to-day for the purpose of determining the latitude and longitude, measuring the height of the walls, drying our rations, and repairing our boats.

Captain Powell early in the morning takes a barometer and goes out to climb a point between the two rivers. I walk down the gorge to the left at the foot of the cliff, climb to a bench, and discover a trail, deeply worn in the rock. Where it crosses the side gulches in some places steps have been cut. I can see no evidence of its having been traveled for a long time. It was doubtless a path used by the people who inhabited this country anterior to the present Indian races–the people who built the communal houses of which mention has been made.

I return to camp about three o'clock and find that some of the men have discovered ruins and many fragments of pottery; also etchings and hieroglyphics on the rocks.

We find to-night, on comparing the readings of the barometers, that the walls are about 3,000 feet high–more than half a mile–an altitude difficult to appreciate from a mere statement of feet. The slope by which the ascent is made is not such a slope as is usually found in climbing a mountain, but one much more abrupt–often vertical for many hundreds of feet,–so that the impression is given that we are at great depths, and we look up to see but a little patch of sky.

Between the two streams, above the Colorado Chiquito, in some places the rocks are broken and shelving for 600 Or 700 feet; then there is a sloping terrace, which can be climbed only by finding some way up a gulch; then another terrace, and back, still another cliff. The summit of the cliff is 3,000 feet above the river, as our barometers attest.

Our camp is below the Colorado Chiquito and on the eastern side of the canyon.

August 12.–The rocks above camp are rust-colored sandstones and conglomerates. Some are very hard; others quite soft. They all lie nearly horizontal, and the beds of softer material have been washed out, leaving the harder forming a series of shelves. Long lines of these are seen, of varying thickness, from one or two to twenty or thirty feet, and the spaces between have the same variability. This morning I spend two or three hours in climbing among these shelves, and then I pass above them and go up a long slope to the foot of the cliff and try to discover some way by which I can reach the top of the wall; but I find my progress cut off by an amphitheater. Then I wander away around to the left, up a little gulch and along benches, climbing from time to time, until I reach an altitude of nearly 2,000 feet and can get no higher. From this point I can look off to the west, up side canyons of the Colorado, and see the edge of a great plateau, from which streams run down into the Colorado, and deep gulches in the escarpment which faces us, continued by canyons, ragged and flaring and set with cliffs and towering crags, down to the river. I can see far up Marble Canyon to long lines of chocolate-colored cliffs, and above these the Vermilion Cliffs. I can see, also, up the Colorado Chiquito, through a very ragged and broken canyon, with sharp salients set out from the walls on either side, their points overlapping, so that a huge tooth of marble on one side seems to be set between two teeth on the opposite; and I can also get glimpses of walls standing away back from the river, while over my head are mural escarpments not possible to be scaled.

Cataract Canyon is 41 miles long. The walls are 1,300 feet high at its head, and they gradually increase in altitude to a point about halfway down, where they are 2,700 feet, and then decrease to 1,300 feet at the foot. Narrow Canyon is 9 1/2 miles long, with walls 1,300 feet in height at the head and coming down to the water at the foot.

There is very little vegetation in this canyon or in the adjacent country. Just at the junction of the Grand and Green there are a number of hackberry trees; and along the entire length of Cataract Canyon the high-water line is marked by scattered trees of the same species. A few nut pines and cedars are found, and occasionally a redbud or Judas tree; but the general aspect of the canyons and of the adjacent country is that of naked rock.

The distance through Glen Canyon is 149 miles. Its walls vary in height from 200 or 300 to 1,600 feet. Marble Canyon is 65 1/2 miles long. At its head it is 200 feet deep, and it steadily increases in depth to its foot, where its walls are 3,500 feet high.

CHAPTER XI. FROM THE LITTLE COLORADO TO THE FOOT OF THE GRAND CANYON.

August 13.–We are now ready to start on our way down the Great Unknown. Our boats, tied to a common, stake, chafe each other as they are tossed by the fretful river. They ride high and buoyant, for their loads are lighter than we could desire. We have but a month's rations remaining. The flour has been resifted through the mosquito-net sieve; the spoiled bacon has been dried and the worst of it boiled; the few pounds of dried apples have been spread in the sun and reshrunken to their normal bulk. The sugar has all melted and gone on its way down the river. But we have a large sack of coffee. The lightening of the boats has this advantage: they will ride the waves better and we shall have but little to carry when we make a portage.

We are three quarters of a mile in the depths of the earth, and the great river shrinks into insignificance as it dashes its angry waves against the walls and cliffs that rise to the world above; the waves are but puny ripples, and we but pigmies, running up and down the sands or lost among the boulders.

We have an unknown distance yet to run, an unknown river to explore. What falls there are, we know not; what rocks beset the channel, we know not; what walls rise over the river, we know not. Ah, well! we may conjecture many things. The men talk as cheerfully as ever; jests are bandied about freely this morning; but to me the cheer is somber and the jests are ghastly.

With some eagerness and some anxiety and some misgiving we enter the canyon below and are carried along by the swift water through walls which rise from its very edge. They have the same structure that we noticed yesterday–tiers of irregular shelves below, and, above these, steep slopes to the foot of marble cliffs. We run six miles in a little more than half an hour and emerge into a more open portion of the canyon, where high hills and ledges of rock intervene between the river and the distant walls. Just at the head of this open place the river runs across a dike; that is, a fissure in the rocks, open to depths below, was filled with eruptive matter, and this on cooling was harder than the rocks through which the crevice was made, and when these were washed away the harder volcanic matter remained as a wall, and the river has cut a gateway through it several hundred feet high and as many wide. As it crosses the wall, there is a fall below and a bad rapid, filled with boulders of trap; so we stop to make a portage. Then on we go, gliding by hills and ledges, with distant walls in view; sweeping past sharp angles of rock; stopping at a few points to examine rapids, which we find can be run, until we have made another five miles, when we land for dinner.

Then we let down with lines over a long rapid and start again. Once more the walls close in, and we find ourselves in a narrow gorge, the water again filling the channel and being very swift. With great care and constant watchfulness we proceed, making about four miles this afternoon, and camp in a cave.

August 14--At daybreak we walk down the bank of the river, on a little sandy beach, to take a view of a new feature in the canyon. Heretofore hard rocks have given us bad river; soft rocks, smooth water; and a series of rocks harder than any we have experienced sets in. The river enters the gneiss! We can see but a little way into the granite gorge, but it looks threatening.

After breakfast we enter on the waves. At the very introduction it inspires awe. The canyon is narrower than we have ever before seen it; the water is swifter; there are but few broken rocks in the channel; but the walls are set, on either side, with pinnacles and crags; and sharp, angular buttresses, bristling with wind- and wave-polished spires, extend far out into the river.

Ledges of rock jut into the stream, their tops sometimes just below the surface, sometimes rising a few or many feet above; and island ledges and island pinnacles and island towers break the swift course of the stream into chutes and eddies and whirlpools. We soon reach a place where a creek comes in from the left, and, just below, the channel is choked with boulders, which have washed down this lateral canyon and formed a dam, over which there is a fall of 30 or

40 feet; but on the boulders foothold can be had, and we make a portage. Three more such dams are found. Over one we make a portage; at the other two are chutes through which we can run.

As we proceed the granite rises higher, until nearly a thousand feet of the lower part of the walls are composed of this rock.

About eleven o'clock we hear a great roar ahead, and approach it very cautiously. The sound grows louder and louder as we run, and at last we find ourselves above a long, broken fall, with ledges and pinnacles of rock obstructing the river. There is a descent of perhaps 75 or 80 feet in a third of a mile, and the rushing waters break into great waves on the rocks, and lash themselves into a mad, white foam. We can land just above, but there is no foothold on either side by which we can make a portage. It is nearly a thousand feet to the top of the granite; so it will be impossible to carry our boats around, though we can climb to the summit up a side gulch and, passing along a mile or two, descend to the river. This we find on examination; but such a portage would be impracticable for us, and we must run the rapid or abandon the river. There is no hesitation. We step into our boats, push off, and away we go, first on smooth but swift water, then we strike a glassy wave and ride to its top, down again into the trough, up again on a higher wave, and down and up on waves higher and still higher until we strike one just as it curls back, and a breaker rolls over our little boat. Still on we speed, shooting past projecting rocks, till the little boat is caught in a whirlpool and spun round several times. At last we pull out again into the stream. And now the other boats have passed us. The open compartment of the "Emma Dean" is filled with water and every breaker rolls over us. Hurled back from a rock, now on this side, now on that, we are carried into an eddy, in which we struggle for a few minutes, and are then out again, the breakers still rolling over us. Our boat is unmanageable, but she cannot sink, and we drift down another hundred yards through breakers–how, we scarcely know. We find the other boats have turned into an eddy at the foot of the fall and are waiting to catch us as we come, for the men have seen that our boat is swamped. They push out as we come near and pull us in against the wall. Our boat bailed, on we go again.

The walls now are more than a mile in height–a vertical distance difficult to appreciate. Stand on the south steps of the Treasury building in Washington and look down Pennsylvania Avenue to the Capitol; measure this distance overhead, and imagine cliffs to extend to that altitude, and you will understand what is meant; or stand at Canal Street in New York and look up Broadway to Grace Church, and you have about the distance; or stand at Lake Street bridge in Chicago and look down to the Central Depot, and you have it again.

A thousand feet of this is up through granite crags; then steep slopes and perpendicular cliffs rise one above another to the summit. The gorge is black and narrow below, red and gray and flaring above, with crags and angular projections on the walls, which, cut in many places by side canyons, seem to be a vast wilderness of rocks. Down in these grand, gloomy depths we glide, ever listening, for the mad waters keep up their roar; ever watching, ever peering ahead, for the narrow canyon is winding and the river is closed in so that we can see but a few hundred yards, and what there may be below we know not; so we listen for falls and watch for rocks, stopping now and then in the bay of a recess to admire the gigantic scenery; and ever as we go there is some new pinnacle or tower, some crag or peak, some distant view of the upper plateau, some strangely shaped rock, or some deep, narrow side canyon.

Then we come to another broken fall, which appears more difficult than the one we ran this morning. A small creek comes in on the right, and the first fall of the water is over boulders, which have been carried down by this lateral stream. We land at its mouth and stop for an hour or two to examine the fall. It seems possible to let down with lines, at least a part of the way, from point to point, along the right-hand wall. So we make a portage over the first rocks and find footing on some boulders below. Then we let down one of the boats to the end of her line, when she reaches a corner of the projecting rock, to which one of the men clings and steadies her while I examine an eddy below. I think we can pass the other boats down by us and catch them in the eddy. This is soon done, and the men in the boats in the eddy pull us to their side. On the shore of this little eddy there is about two feet of gravel beach above the water. Standing on this beach, some of the men take the line of the little boat and let it drift down against another projecting angle. Here is a little shelf, on which a man from my boat climbs, and a shorter line is passed to him, and he fastens the boat to the side of the cliff; then the second one is let down, bringing the line of the third. When the second boat is tied up, the two men standing on the beach above spring into the last boat, which is pulled up alongside of ours; then we let down the boats for 25 or 30 yards by walking along the shelf, landing them again in the mouth of a side canyon. Just below this there is another pile of boulders, over which we make another portage. From the foot of these rocks we can climb to another shelf, 40 or 50 feet above the water.

On this bench we camp for the night. It is raining hard, and we have no shelter, but find a few sticks which have lodged in the rocks, and kindle a fire and have supper. We sit on the rocks all night, wrapped in our ponchos, getting what sleep we can.

August 15.–This morning we find we can let down for 300 or 400 yards, and it is managed in this way: we pass along the wall by climbing from projecting point to point, sometimes near the water's edge, at other places 50 or 60 feet above, and hold the boat with a line while two men remain aboard and prevent her from being dashed against the rocks and keep the line from getting caught on the wall. In two hours we have brought them all down, as far as it is possible, in this way. A few yards below, the river strikes with great violence against a projecting rock and our boats are pulled up in a little bay above. We must now manage to pull out of this and clear the point below. The little boat is held by the bow obliquely up the stream. We jump in and pull out only a few strokes, and sweep clear of the dangerous rock. The other boats follow in the same manner and the rapid is passed.

It is not easy to describe the labor of such navigation. We must prevent the waves from dashing the boats against the cliffs. Sometimes, where the river is swift, we must put a bight of rope about a rock, to prevent the boat from being snatched from us by a wave; but where the plunge is too great or the chute too swift, we must let her leap and catch her below or the undertow will drag her under the falling water and sink her. Where we wish to run her out a little way from shore through a channel between rocks, we first throw in little sticks of driftwood and watch their course, to see where we must steer so that she will pass the channel in safety. And so we hold, and let go, and pull, and lift, and ward–among rocks, around rocks, and over rocks.

And now we go on through this solemn, mysterious way. The river is very deep, the canyon very narrow, and still obstructed, so that there is no steady flow of the stream; but the waters reel and roll and boil, and we are scarcely able to determine where we can go. Now the boat is carried to the right, perhaps close to the wall; again, she is shot into the stream, and perhaps is dragged over to the other side, where, caught in a whirlpool, she spins about. We can neither land nor run as we please. The boats are entirely unmanageable; no order in their running can be preserved; now one, now another, is ahead, each crew laboring for its own preservation. In such a place we come to another rapid. Two of the boats run it perforce. One succeeds in landing, but there is no foothold by which to make a portage and she is pushed out again into the stream. The next minute a great reflex wave fills the open compartment; she is water-logged, and drifts unmanageable. Breaker after breaker rolls over her and one capsizes her. The men are thrown out; but they cling to the boat, and she drifts down some distance alongside of us and we are able to catch her. She is soon bailed out and the men are aboard once more; but the oars are lost, and so a pair from the "Emma Dean" is spared. Then for two miles we find smooth water.

Clouds are playing in the canyon to-day. Sometimes they roll down in great masses, filling the gorge with gloom; sometimes they hang aloft from wall to wall and cover the canyon with a roof of impending storm, and we can peer long distances up and down this canyon corridor, with its cloud-roof overhead, its walls of black granite, and its river bright with the sheen of broken waters. Then a gust of wind sweeps down a side gulch and, making a rift in the clouds, reveals the blue heavens, and a stream of sunlight pours in. Then the clouds drift away into the distance, and hang around crags and peaks and pinnacles and towers and walls, and cover them with a mantle that lifts from time to time and sets them all in sharp relief. Then baby clouds creep out of side canyons, glide around points, and creep back again into more distant gorges. Then clouds arrange in strata across the canyon, with intervening vista views to cliffs and rocks beyond. The clouds are children of the heavens, and when they play among the rocks they lift them to the region above.

It rains! Rapidly little rills are formed above, and these soon grow into brooks, and the brooks grow into creeks and tumble over the walls in innumerable cascades, adding their wild music to the roar of the river. When the rain ceases the rills, brooks, and creeks run dry. The waters that fall during a rain on these steep rocks are gathered at once into the river; they could scarcely be poured in more suddenly if some vast spout ran from the clouds to the stream itself. When a storm bursts over the canyon a side gulch is dangerous, for a sudden flood may come, and the inpouring waters will raise the river so as to hide the rocks.

Early in the afternoon we discover a stream entering from the north–a clear, beautiful creek, coming down through a gorgeous red canyon. We land and camp on a sand beach above its mouth, under a great, overspreading tree with willow-shaped leaves.

August 16.–We must dry our rations again to-day and make oars.

The Colorado is never a clear stream, but for the past three or four days it has been raining much of the time, and the floods poured over the walls have brought down great quantities of mud, making it exceedingly turbid now. The little affluent which we have discovered here is a clear, beautiful creek, or river, as it would be termed in this western country, where streams are not abundant. We have named one stream, away above, in honor of the great chief of the "Bad Angels," and as this is in beautiful contrast to that, we conclude to name it "Bright Angel."

Early in the morning the whole party starts up to explore the Bright Angel River, with the special purpose of seeking timber from which to make oars. A couple of miles above we find a large pine log, which has been floated down from the plateau, probably from an altitude of more than 6,000 feet, but not many miles back. On its way it must have passed over many cataracts and falls, for it bears scars in evidence of the rough usage which it has received. The men roll it on skids, and the work of sawing oars is commenced.

This stream heads away back under a line of abrupt cliffs that terminates the plateau, and tumbles down more than 4,000 feet in the first mile or two of its course; then runs through a deep, narrow canyon until it reaches the river.

Late in the afternoon I return and go up a little gulch just above this creek, about 200 yards from camp, and discover the ruins of two or three old houses, which were originally of stone laid in mortar. Only the foundations are left, but irregular blocks, of which the houses were constructed, lie scattered about. In one room I find an old mealing-stone, deeply worn, as if it had been much used. A great deal of pottery is strewn around, and old trails, which in some places are deeply worn into the rocks, are seen.

It is ever a source of wonder to us why these ancient people sought such inaccessible places for their homes. They were, doubtless, an agricultural race, but there are no lands here of any considerable extent that they could have cultivated. To the west of Oraibi, one of the towns in the Province of Tusayan, in northern Arizona, the inhabitants have actually built little terraces along the face of the cliff where a spring gushes out, and thus made their sites for gardens. It is possible that the ancient inhabitants of this place made their agricultural lands in the same way. But why should they seek such spots'? Surely the country was not so crowded with people as to demand the utilization of so barren a region. The only solution suggested of the problem is this: We know that for a century or two after the settlement of Mexico many expeditious were sent into the country now comprising Arizona and New Mexico, for the purpose of bringing the town-building people under the dominion of the Spanish government. Many of their villages were destroyed, and the inhabitants fled to regions at that time unknown; and there are traditions among the people who inhabit the pueblos that still remain that the canyons were these unknown lauds. It may be these buildings were erected at that time; sure it is that they have a much more modern appearance than the ruins scattered over Nevada, Utah, Colorado, Arizona, and New Mexico. Those old Spanish conquerors had a monstrous greed for gold and a wonderful lust for saving souls. Treasures they must have, if not on earth, why, then, in heaven; and when they failed to find heathen temples bedecked with silver, they propitiated Heaven by seizing the heathen themselves. There is yet extant a copy of a record made by a heathen artist to express his conception of the demands of the conquerors. In one part of the picture we have a lake, and near by stands a priest pouring water on the head of a native. On the other side, a poor Indian has a cord about his throat. Lines run from these two groups to a central figure, a man with beard and full Spanish panoply. The interpretation of the picture-writing is this: "Be baptized as this saved heathen, or be hanged as that damned heathen." Doubtless, some of these people preferred another alternative, and rather than be baptized or hanged they chose to imprison themselves within these canyon walls.

August 17.–Our rations are still spoiling; the bacon is so badly injured that we are compelled to throw it away. By an accident, this morning, the saleratus was lost overboard. We have now only musty flour sufficient for ten days and a few dried apples, but plenty of coffee. We must make all haste possible. If we meet with difficulties such as we have encountered in the canyon above, we may be compelled to give up the expedition and try to reach the Mormon settlements to the north.

Our hopes are that the worst places are passed, but our barometers are all so much injured as to be useless, and so we have lost our reckoning in altitude, and know not how much descent the river has yet to make. The stream is still wild and rapid and rolls through a narrow channel. We make but slow progress, often landing against a wall and climbing around some point to see the river below. Although very anxious to advance, we are determined to run with great caution, lest by another accident we lose our remaining supplies. How precious that little flour has become! We divide it among the boats and carefully store it away, so that it can be lost only by the loss of the boat itself.

We make ten miles and a half, and camp among the rocks on the right. We have had rain from time to time all day, and have been thoroughly drenched and chilled; but between showers the sun shines with great power and the mercury in our thermometers stands at 115 degrees, so that we have rapid changes from great extremes, which are very disagreeable. It is especially cold in the rain to-night. The little canvas we have is rotten and useless; the rubber ponchos with which we started from Green River City have all been lost; more than half the party are without hats, not one of us has an entire suit of clothes, and we have not a blanket apiece. So we gather driftwood and build a fire; but after supper the rain, coming down in torrents, extinguishes it, and we sit up all night on the rocks, shivering, and are more exhausted by the night's discomfort than by the day's toil.

August 18.–The day is employed in making portages and we advance but two miles on our journey. Still it rains.

While the men are at work making portages I climb up the granite to its summit and go away back over the rust-colored sandstones and greenish-yellow shales to the foot of the marble wall. I climb so high that the men and boats are lost in the black depths below and the dashing river is a rippling brook, and still there is more canyon above than below. All about me are interesting geologic records. The book is open and I can read as I run. All about me are grand views, too, for the clouds are playing again in the gorges. But somehow I think of the nine days' rations and the bad river, and the lesson of the rocks and the glory of the scene are but half conceived. I push on to an angle, where I hope to get a view of the country beyond, to see if possible what the prospect may be of our soon running through this plateau, or at least of meeting with some geologic change that will let us out of the granite; but, arriving at the point, I can see below only a labyrinth of black gorges.

August 19.–Rain again this morning. We are in our granite prison still, and the time until noon is occupied in making a long; bad portage.

After dinner, in running a rapid the pioneer boat is upset by a wave. We are some distance in advance of the larger boats. The river is rough and swift and we are unable to land, but cling to the boat and are carried down stream over another rapid. The men in the boats above see our trouble, but they are caught in whirlpools and are spinning about in eddies, and it seems a long time before they come to our relief. At last they do come; our boat is turned right side up and bailed out; the oars, which fortunately have floated along in company with us, are gathered up, and on we go, without even landing. The clouds break away and we have sunshine again.

Soon we find a little beach with just room enough to land. Here we camp, but there is no wood. Across the river and a little way above, we see some driftwood lodged in the rocks. So we bring two boat loads over, build a huge fire, and spread everything to dry. It is the first cheerful night we have had for a week–a warm, drying fire in the midst of the camp, and a few bright stars in our patch of heavens overhead.

August 20.–The characteristics of the canyon change this morning. The river is broader, the walls more sloping, and composed of black slates that stand on edge. These nearly vertical slates are washed out in places–that is, the softer beds are washed out between the harder, which are left standing. In this way curious little alcoves are formed, in which are quiet bays of water, but on a much smaller scale than the great bays and buttresses of Marble Canyon.

The river is still rapid and we stop to let down with lines several times, but make greater progress, as we run ten miles. We camp on the right bank. Here, on a terrace of trap, we discover another group of ruins. There was evidently quite a village on this rock. Again we find mealing-stones and much broken pottery, and up on a little natural shelf in the rock back of the ruins we find a globular basket that would hold perhaps a third of a bushel. It is badly broken, and as I attempt to take it up it falls to pieces. There are many beautiful flint chips, also, as if this had been the home of an old arrow-maker.

August 21.–We start early this morning, cheered by the prospect of a fine day and encouraged also by the good run made yesterday. A quarter of a mile below camp the river turns abruptly to the left, and between camp and that point is very swift, running down in a long, broken chute and piling up against the foot of the cliff, where it turns to the left. We try to pull across, so as to go down on the other side, but the waters are swift and it seems impossible for us to escape the rock below; but, in pulling across, the bow of the boat is turned to the farther shore, so that we are swept broadside down and are prevented by the rebounding waters from striking against the wall. We toss about for a few seconds in these billows and are then carried past the danger. Below, the river turns again to the right, the canyon is very narrow, and we see in advance but a short distance. The water, too, is very swift, and there is no landing-place. From around this curve there comes a mad roar, and down we are carried with a dizzying velocity to the head of another rapid. On either side high over our heads there are overhanging granite walls, and the sharp bends cut off our view, so that a few minutes will carry us into unknown waters. Away we go on one long, winding chute. I stand on deck, supporting myself with a strap fastened on either side of the gunwale. The boat glides rapidly where the water is smooth, then, striking a wave, she leaps and bounds like a thing of life, and we have a wild, exhilarating ride for ten miles, which we make in less than an hour. The excitement is so great that we forget the danger until we hear the roar of a great fall below; then we back on our oars and are carried slowly toward its head and succeed in landing just above and find that we have to make another portage. At this we are engaged until some time after dinner.

Just here we run out of the granite. Ten miles in less than half a day, and limestone walls below. Good cheer returns; we forget the storms and the gloom and the cloud-covered canyons and the black granite and the raging river, and push our boats from shore in great glee.

Though we are out of the granite, the river is still swift, and we wheel about a point again to the right, and turn, so as to head back in the direction from which we came; this brings the granite in sight again, with its narrow gorge and black crags; but we meet with no more great falls or rapids. Still, we run cautiously and stop from time to time to examine some places which look bad. Yet we make ten miles this afternoon; twenty miles in all to-day.

August 22.–We come to rapids again this morning and are occupied several hours in passing them, letting the boats down from rock to rock with lines for nearly half a mile, and then have to make a long portage. While the men are engaged in this I climb the wall on the northeast to a height of about 2,500 feet, where I can obtain a good view of a long stretch of canyon below. Its course is to the southwest. The walls seem to rise very abruptly for 2,500 or 3,000 feet, and then there is a gently sloping terrace on each side for two or three miles, when we again find cliffs, 1,500 or 2,000 feet high. From the brink of these the plateau stretches back to the north and south for a long distance. Away down the canyon on the right wall I can see a group of mountains, some of which appear to stand on the brink of the canyon. The effect of the terrace is to give the appearance of a narrow winding valley with high walls on either side and a deep, dark, meandering gorge down its middle. It is impossible from this point of view to determine whether or not we have granite at the bottom; but from geologic considerations, I conclude that we shall have marble walls below.

After my return to the boats we run another mile and camp for the night. We have made but little over seven miles to-day, and a part of our flour has been soaked in the river again.

August 23.–Our way to-day is again through marble walls. Now and then we pass for a short distance through patches of granite, like hills thrust up into the limestone. At one of these places we have to make another portage, and, taking advantage of the delay, I go up a little stream to the north, wading it all the way, sometimes having to plunge in to my neck, in other places being compelled to swim across little basins that have been excavated at the foot of the falls. Along its course are many cascades and springs, gushing out from the rocks on either side. Sometimes a cottonwood tree grows over the water. I come to one beautiful fall, of more than 150 feet, and climb around it to the right on the broken rocks. Still going up, the canyon is found to narrow very much, being but 15 or 20 feet wide; yet the walls rise on either side many hundreds of feet, perhaps thousands; I can hardly tell.

In some places the stream has not excavated its channel down vertically through the rocks, but has cut obliquely, so that one wall overhangs the other. In other places it is cut vertically above and obliquely below, or obliquely above and vertically below, so that it is impossible to see out overhead. But I can go no farther; the time which I estimated it would take to make the portage has almost expired, and I start back on a round trot, wading in the creek where I must and plunging through basins. The men are waiting for me, and away we go on the river.

Just after dinner we pass a stream on the right, which leaps into' the Colorado by a direct fall of more than 100 feet, forming a beautiful cascade. There is a bed of very hard rock above, 30 or 40 feet in thickness, and there are much softer beds below. The hard beds above project many yards beyond the softer, which are washed out, forming a deep cave behind the fall, and the stream pours through a narrow crevice above into a deep pool below. Around on the rocks in the cavelike chamber are set beautiful ferns, with delicate fronds and enameled stalks. The frondlets have their points turned down to form spore cases. It has very much the appearance of the maidenhair fern, but is much larger. This delicate foliage covers the rocks all about the fountain, and gives the chamber great beauty. But we have little time to spend in admiration; so on we go.

We make fine progress this afternoon, carried along by a swift river, shooting over the rapids and finding no serious obstructions. The canyon walls for 2,500 or 3,000 feet are very regular, rising almost perpendicularly, but here and there set with narrow steps, and occasionally we can see away above the broad terrace to distant cliffs.

We camp to-night in a marble cave, and find on looking at our reckoning that we have run 22 miles.

August 24.–The canyon is wider to-day. The walls rise to a vertical height of nearly 3,000 feet. In many places the river runs under a cliff in great curves, forming amphitheaters half-dome shaped.

Though the river is rapid, we meet with no serious obstructions and run 20 miles. How anxious we are to make up our reckoning every time we stop, now that our diet is confined to plenty of coffee, a very little spoiled flour, and very few dried apples! It has come to be a race for a dinner. Still, we make such fine progress that all hands are in good cheer, but not a moment of daylight is lost.

August 25.–We make 12 miles this morning, when we come to monuments of lava standing in the river,–low rocks mostly, but some of them shafts more than a hundred feet high. Going on down three or four miles, we find them increasing in number. Great quantities of cooled lava and many cinder cones are seen on either side; and then we come to an abrupt cataract. Just over the fall on the right wall a cinder cone, or extinct volcano, with a well-defined crater, stands on the very brink of the canyon. This, doubtless, is the one we saw two or three days ago. From this volcano vast floods of lava have been poured down into the river, and a stream of molten rock has run up the canyon three or four miles and down we know not how far. Just where it poured over the canyon wall is the fall. The whole north side as far as we can see is lined with the black basalt, and high up on the opposite wall are patches of the same material, resting on the benches and filling old alcoves and caves, giving the wall a spotted appearance.

The rocks are broken in two along a line which here crosses the river, and the beds we have seen while coming down the canyon for the last 30 miles have dropped 800 feet on the lower side of the line, forming what geologists call a "fault." The volcanic cone stands directly over the fissure thus formed. On the left side of the river, opposite, mammoth springs burst out of this crevice, 100 or 200 feet above the river, pouring in a stream quite equal in volume to the Colorado Chiquito.

This stream seems to be loaded with carbonate of lime, and the water, evaporating, leaves an incrustation on the rocks; and this process has been continued for a long time, for extensive deposits are noticed in which are basins with bubbling springs. The water is salty.

We have to make a portage here, which is completed in about three hours; then on we go.

We have no difficulty as we float along, and I am able to observe the wonderful phenomena connected with this flood of lava. The canyon was doubtless filled to a height of 1,200 or 1,500 feet, perhaps by more than one flood. This would dam the water back; and in cutting through this great lava bed, a new channel has been formed, sometimes on one side, sometimes on the other. The cooled lava, being of firmer texture than the rocks of which the walls are composed, remains in some places; in others a narrow channel has been cut, leaving a line of basalt on either side. It is possible that the lava cooled faster on the sides against the walls and that the center ran out; but of this we can only conjecture. There are other places where almost the whole of the lava is gone, only patches of it being seen, where it has caught on the walls. As we float down we can see that it ran out into side canyons. In some places this basalt has a fine, columnar structure, often in concentric prisms, and masses of these concentric columns have coalesced. In some places, when the flow occurred the canyon was probably about the same depth that it is now, for we can see where the basalt has rolled out on the sands, and–what seems curious to me–the sands are not melted or metamorphosed to any appreciable extent. In places the bed of the river is of sandstone or limestone, in other places of lava, showing that it has all been cut out again where the sandstones and limestones appear; but there is a little yet left where the bed is of lava.

What a conflict of water and fire there must have been here! Just imagine a river of molten rock running down into a river of melted snow. What a seething and boiling of the waters; what clouds of steam rolled into the heavens!

Thirty-five miles to-day. Hurrah!

August 26.–The canyon walls are steadily becoming higher as we advance. They are still bold and nearly vertical up to the terrace. We still see evidence of the eruption discovered yesterday, but the thickness of the basalt is decreasing as we go down stream; yet it has been reinforced at points by streams that have come down from volcanoes standing on the terrace above, but which we cannot see from the river below.

Since we left the Colorado Chiquito we have seen no evidences that the tribe of Indians inhabiting the plateaus on either side ever come down to the river; but about eleven o'clock to-day we discover an Indian garden at the foot of the wall on the right, just where a little stream with a narrow flood plain comes down through a side canyon. Along the valley the Indians have planted corn, using for irrigation the water which bursts out in springs at the foot of the cliff. The corn is looking quite well, but it is not sufficiently advanced to give us roasting ears; but there are some nice green squashes. We carry ten or a dozen of these on board our boats and hurriedly leave, not willing to be caught in the robbery, yet excusing ourselves by pleading our great want. We run down a short distance to where we feel certain no Indian can follow, and what a kettle of squash sauce we make! True, we have no salt with which to season it, but it makes a fine addition to our unleavened bread and coffee. Never was fruit so sweet as these stolen squashes.

After dinner we push on again and make fine time, finding many rapids, but none so bad that we cannot run them with safety; and when we stop, just at dusk, and foot up our reckoning, we find we have run 35 miles again. A few days like this, and we are out of prison.

We have a royal supper–unleavened bread, green squash sauce, and strong coffee. We have been for a few days on half rations, but now have no stint of roast squash.

August 27.–This morning the river takes a more southerly direction. The dip of the rocks is to the north and we are running rapidly into lower formations. Unless our course changes we shall very soon run again into the granite. This gives some anxiety. Now and then the river turns to the west and excites hopes that are soon destroyed by another turn to the south. About nine o'clock we come to the dreaded rock. It is with no little misgiving that we see the river enter these black, hard walls. At its very entrance we have to make a portage; then let down with lines past some ugly rocks. We run a mile or two farther, and then the rapids below can be seen.

About eleven o'clock we come to a place in the river which seems much worse than any we have yet met in all its course. A little creek comes down from the left. We land first on the right and clamber up over the granite pinnacles for a mile or two, but can see no way by which to let down, and to run it would be sure destruction. After dinner we cross to examine on the left. High above the river we can walk along on the top of the granite, which is broken off at the edge and set with crags and pinnacles, so that it is very difficult to get a view of the river at all. In my eagerness to reach a point where I can see the roaring fall below, I go too far on the wall, and can neither advance nor retreat. I stand with one foot on a little projecting rock and cling with my hand fixed in a little crevice. Finding I am caught here, suspended 400 feet above the river, into which I must fall if my footing fails, I call for help. The men come and pass me a line, but I cannot let go of the rock long enough to take hold of it. Then they bring two or three of the largest oars. All this takes time which seems very precious to me; but at last they arrive. The blade of one of the oars is pushed into a little crevice in the rock beyond me in such a manner that they can hold me pressed against the wall. Then another is fixed in such a way that I can step on it; and thus I am extricated.

Still another hour is spent in examining the river from this side, but no good view of it is obtained; so now we return to the side that was first examined, and the afternoon is spent in clambering among the crags and pinnacles and carefully scanning the river again. We find that the lateral streams have washed boulders into the river, so as to form a dam, over which the water makes a broken fall of 18 or 20 feet; then there is a rapid, beset with rocks, for 200 or 300 yards, while on the other side, points of the wall project into the river. Below, there is a second fall; how great, we cannot tell. Then there is a rapid, filled with huge rocks, for 100 or 200 yards. At the bottom of it, from the right wall, a great rock projects quite halfway across the river. It has a sloping surface extending up stream, and the water, coming down with all the momentum gained in the falls and rapids above, rolls up this inclined plane many feet, and tumbles over to the left. I decide that it is possible to let down over the first fall, then run near the right cliff to a point just above the second, where we can pull out into a little chute, and, having run over that in safety, if we pull with all our power across the stream, we may avoid the great rock below. On my return to the boat I announce to the men that we are to run it in the morning. Then we cross the river and go into camp for the night on some rocks in the mouth of the little side canyon.

After supper Captain Howland asks to have a talk with me. We walk up the little creek a short distance, and I soon find that his object is to remonstrate against my determination to proceed. He thinks that we had better abandon the river here. Talking with him, I learn that he, his brother, and William Dunn have determined to go no farther in the boats. So we return to camp. Nothing is said to the other men.

For the last two days our course has not been plotted. I sit down and do this now, for the purpose of finding where we are by dead reckoning. It is a clear night, and I take out the sextant to make observation for latitude, and I find that the astronomic determination agrees very nearly with that of the plot–quite as closely as might be expected from a meridian observation on a planet. In a direct line, we must be about 45 miles from the mouth of the Rio Virgen. If we can reach that point, we know that there are settlements up that river about 20 miles. This 45 miles in a direct line will probably be 80 or 90 by the meandering line of the river. But then we know that there is comparatively open country for many miles above the mouth of the Virgen, which is our point of destination.

As soon as I determine all this, I spread my plot on the sand and wake Howland, who is sleeping down by the river, and show him where I suppose we are, and where several Mormon settlements are situated.

We have another short talk about the morrow, and he lies down again; but for me there is no sleep. All night long I pace up and down a little path, on a few yards of sand beach, along by the river. Is it wise to go on? I go to the boats again to look at our rations. I feel satisfied that we can get over the danger immediately before us; what there may be below I know not. From our outlook yesterday on the cliffs, the canyon seemed to make another great bend to the south, and this, from our experience heretofore, means more and higher granite walls. I am not sure that we can climb out of the canyon here, and, if at the top of the wall, I know enough of the country to be certain that it is a desert of rock and sand between this and the nearest Mormon town, which, on the most direct line, must be 75 miles away. True, the late rains have been favorable to us, should we go out, for the probabilities are that we shall find water still standing in holes; and at one time I almost conclude to leave the river. But for years I have been contemplating this trip. To leave the exploration unfinished, to say that there is a part of the canyon which I cannot explore, having already nearly accomplished it, is more than I am willing to acknowledge, and I determine to go on.

I wake my brother and tell him of Howland's determination, and he promises to stay with me; then I call up Hawkins, the cook, and he makes a like promise; then Sumner and Bradley and Hall, and they all agree to go on.

August 28.–At last daylight comes and we have breakfast without a word being said about the future. The meal is as solemn as a funeral. After breakfast I ask the three men if they still think it best to leave us. The elder Howland thinks it is, and Dunn agrees with him. The younger Howland tries to persuade them to go on with the party; failing in which, he decides to go with his brother.

Then we cross the river. The small boat is very much disabled and unseaworthy. With the loss of hands, consequent on the departure of the three men, we shall not be able to run all of the boats; so I decide to leave my "Emma Dean."

Two rifles and a shotgun are given to the men who are going out. I ask them to help themselves to the rations and take what they think to be a fair share. This they refuse to do, saying they have no fear but that they can get something to eat; but Billy, the cook, has a pan of biscuits prepared for dinner, and these he leaves on a rock.

Before starting, we take from the boat our barometers, fossils, the minerals, and some ammunition and leave them on the rocks. We are going over this place as light as possible. The three men help us lift our boats over a rock 25 or 30 feet high and let them down again over the first fall, and now we are all ready to start. The last thing before leaving, I write a letter to my wife and give it to Howland. Sumner gives him his watch, directing that it be sent to his sister should he not be heard from again. The records of the expedition have been kept in duplicate. One set of these is given to Howland; and now we are ready. For the last time they entreat us not to go on, and tell us that it is madness to set out in this place; that we can never get safely through it; and, further, that the river turns again to the south into the granite, and a few miles of such rapids and falls will exhaust our entire stock of rations, and then it will be too late to climb out. Some tears are shed; it is rather a solemn parting; each party thinks the other is taking the dangerous course.

My old boat left, I go on board of the "Maid of the Canyon." The three men climb a crag that overhangs the river to watch us off. The "Maid of the Canyon" pushes out. We glide rapidly along the foot of the wall, just grazing one great rock, then pull out a little into the chute of the second fall and plunge over it. The open compartment is filled when we strike the first wave below, but we cut through it, and then the men pull with all their power toward the left wall and swing clear of the dangerous rock below all right. We are scarcely a minute in running it, and find that, although it looked bad from above, we have passed many places that were worse. The other boat follows without more difficulty. We land at the first practicable point below, and fire our guns, as a signal to the men above that we have come over in safety. Here we remain a couple of hours, hoping that they will take the smaller boat and follow us. We are behind a curve in the canyon and cannot see up to where we left them, and so we wait until their coming seems hopeless, and then push on.

And now we have a succession of rapids and falls until noon, all of which we run in safety. Just after dinner we come to another bad place. A little stream comes in from the left, and below there is a fall, and still below another fall. Above, the river tumbles down, over and among the rocks, in whirlpools and great waves, and the waters are lashed into mad, white foam. We run along the left, above this, and soon see that we cannot get down on this side, but it seems possible to let down on the other. We pull up stream again for 200 or 300 yards and cross. Now there is a bed of basalt on this northern side of the canyon, with a bold escarpment that seems to be a hundred feet high. We can climb it and walk along its summit to a point where we are just at the head of the fall. Here the basalt is broken down again, so it seems to us, and I direct the men to take a line to the top of the cliff and let the boats down along the wall. One man remains in the boat to keep her clear of the rocks and prevent her line from being caught on the projecting angles. I climb the cliff and pass along to a point just over the fall and descend by broken rocks, and find that the break of the fall is above the break of the wall, so that we cannot land, and that still below the river is very bad, and that there is no possibility of a portage. Without waiting further to examine and determine what shall be done, I hasten back to the top of the cliff to stop the boats from coming down. When I arrive I find the men have let one of them down to the head of the fall. She is in swift water and they are not able to pull her back; nor are they able to go on with the line, as it is not long enough to reach the higher part of the cliff which is just before them; so they take a bight around a crag. I send two men back for the other line. The boat is in very swift water, and Bradley is standing in the open compartment, holding out his oar to prevent her from striking against the foot of the cliff. Now she shoots out into the stream and up as far as the line will permit, and then, wheeling, drives headlong against the rock, and then out and back again, now straining on the line, now striking against the rock. As soon as the second line is brought, we pass it down to him; but his attention is all taken up with his own situation, and he does not see that we are passing him the line. I stand on a projecting rock, waving my hat to gain his attention, for my voice is drowned by the roaring of the falls. Just at this moment I see him take his knife from its sheath and step forward to cut the line. He has evidently decided that it is better to go over with the boat as it is than to wait for her to be broken to pieces. As he leans over, the boat sheers again into the stream, the stem-post breaks away and she is loose. With perfect composure Bradley seizes the great scull oar, places it in the stern rowlock, and pulls with all his power (and he is an athlete) to turn the bow of the boat down stream, for he wishes to go bow down, rather than to drift broadside on. One, two strokes he makes, and a third just as she goes over, and the boat is fairly turned, and she goes down almost beyond our sight, though we are more than a hundred feet above the river. Then she comes up again on a great wave, and down and up, then around behind some great rocks, and is lost in the mad, white foam below. We stand frozen with fear, for we see no boat. Bradley is gone! so it seems. But now, away below, we see something coming out of the waves. It is evidently a boat. A moment more, and we see Bradley standing on deck, swinging his hat to show that he is all right. But he is in a whirlpool. We have the stem-post of his boat attached to the line. How badly she may be disabled we know not. I direct Sumner and Powell to pass along the cliff and see if they can reach him from below. Hawkins, Hall, and myself run to the other boat, jump aboard, push out, and away we go over the falls. A wave rolls over us and our boat is unmanageable. Another great wave strikes us, and the boat rolls over, and tumbles and tosses, I know not how. All I know is that Bradley is picking us up. We soon have all right again, and row to the cliff and wait until Sumner and Powell can come. After a difficult climb they reach us. We run two or three miles farther and turn again to the northwest, continuing until night, when we have run out of the granite once more.

August 29.–We start very early this morning. The river still continues swift, but we have no serious difficulty, and at twelve o'clock emerge from the Grand Canyon of the Colorado. We are in a valley now, and low mountains are seen in the distance, coming to the river below. We recognize this as the Grand Wash.

A few years ago a party of Mormons set out from St. George, Utah, taking with them a boat, and came down to the Grand Wash, where they divided, a portion of the party crossing the river to explore the San Francisco Mountains. Three men–Hamblin, Miller, and Crosby–taking the boat, went on down the river to Callville, landing a few miles below the mouth of the Rio Virgen. We have their manuscript journal with us, and so the stream is comparatively well known.

To-night we camp on the left bank, in a mesquite thicket.

The relief from danger and the joy of success are great. When he who has been chained by wounds to a hospital cot until his canvas tent seems like a dungeon cell, until the groans of those who lie about tortured with probe and knife are piled up, a weight of horror on his ears that he cannot throw off, cannot forget, and until the stench of festering wounds and anaesthetic drugs has filled the air with its loathsome burthen,–when he at last goes out into the open field, what a world he sees! How beautiful the sky, how bright the sunshine, what "floods of delirious music" pour from the throats of birds, how sweet the fragrance of earth and tree and blossom! The first hour of convalescent freedom seems rich recompense for all pain and gloom and terror.

Something like these are the feelings we experience to-night. Ever before us has been an unknown danger, heavier than immediate peril. Every waking hour passed in the Grand Canyon has been one of toil. We have watched with deep solicitude the steady disappearance of our scant supply of rations, and from time to time have seen the river snatch a portion of the little left, while we were a-hungered. And danger and toil were endured in those gloomy depths, where ofttimes clouds hid the sky by day and but a narrow zone of stars could be seen at night. Only during the few hours of deep sleep, consequent on hard labor, has the roar of the waters been hushed. Now the danger is over, now the toil has ceased, now the gloom has disappeared, now the firmament is bounded only by the horizon, and what a vast expanse of constellations can be seen!

The river rolls by us in silent majesty; the quiet of the camp is sweet; our joy is almost ecstasy. We sit till long after midnight talking of the Grand Canyon, talking of home, but talking chiefly of the three men who left us. Are they wandering in those depths, unable to find a way out? Are they searching over the desert lands above for water? Or are they nearing the settlements?

August 30.–We run in two or three short, low canyons to-day, and on emerging from one we discover a band of Indians in the valley below. They see us, and scamper away in eager haste to hide among the rocks. Although we land and call for them to return, not an Indian can be seen.

Two or three miles farther down, in turning a short bend of the river, we come upon another camp. So near are we before they can see us that I can shout to them, and, being able to speak a little of their language, I tell them we are friends; but they all flee to the rocks, except a man, a woman, and two children. We land and talk with them. They are without lodges, but have built little shelters of boughs, under which' they wallow in the sand. The man is dressed in a hat; the woman, in a string of beads only. At first they are evidently much terrified; but when I talk to them in their own language and tell them we are friends, and inquire after people in the Mormon towns, they are soon reassured and beg for tobacco. Of this precious article we have none to spare. Sumner looks around in the boat for something to give them, and finds a little piece of colored soap, which they receive as a valuable present,–rather as a thing of beauty than as a useful commodity, however. They are either unwilling or unable to tell us anything about the Indians or white people, and so we push off, for we must lose no time.

We camp at noon under the right bank. And now as we push out we are in great expectancy, for we hope every minute to discover the mouth of the Rio Virgen. Soon one of the men exclaims: "Yonder's an Indian in the river." Looking for a few minutes, we certainly do see two or three persons. The men bend to their oars and pull toward them. Approaching, we see that there are three white men and an Indian hauling a seine, and then we discover that it is just at the mouth of the long-sought river.

As we come near, the men seem far less surprised to see us than we do to see them. They evidently know who we are, and on talking with them they tell us that we have been reported lost long ago, and that some weeks before a messenger had been sent from Salt Lake City with instructions for them to watch for any fragments or relics of our party that might drift down the stream.

Our new-found friends, Mr. Asa and his two sons, tell us that they are pioneers of a town that is to be built on the bank. Eighteen or twenty miles up the valley of the Rio Virgen there are two Mormon towns, St. Joseph and St. Thomas. To-night we dispatch an Indian to the last-mentioned place to bring any letters that may be there for us.

Our arrival here is very opportune. When we look over our store of supplies, we find about 10 pounds of flour, 15 pounds of dried apples, but 70 or 80 pounds of coffee.

August 81.–This afternoon the Indian returns with a letter informing us that Bishop Leithhead of St. Thomas and two or three other Mormons are coming down with a wagon, bringing us supplies. They arrive about sundown. Mr. Asa treats us with great kindness to the extent of his ability; but Bishop Leithhead brings in his wagon two or three dozen melons and many other little luxuries, and we are comfortable once more.

September 1.–This morning Sumner, Bradley, Hawkins, and Hall, taking on a small supply of rations, start down the Colorado with the boats. It is their intention to go to Fort Mojave, and perhaps from there overland to Los Angeles.

Captain Powell and myself return with Bishop Leithhead to St. Thomas. From St. Thomas we go to Salt Lake City.

CHAPTER XII. THE RIO VIRGEN AND THE UINKARET MOUNTAINS.

A year has passed, and we have determined to resume the exploration of the canyons of the Colorado. Our last trip was so hurried, owing to the loss of rations, and the scientific instruments were so badly injured, that we are not satisfied with the results obtained; so we shall once more attempt to pass through the canyons in boats, devoting two or three years to the trip.

It will not be possible to carry in the boats sufficient supplies for the party for that length of time; so it is thought best to establish depots of supplies, at intervals of 100 or 200 miles along the river.

Between Gunnison's Crossing and the foot of the Grand Canyon, we know of only two points where the river can be reached–one at the Crossing of the Fathers, and another a few miles below, at the mouth of the Paria, on a route which has been explored by Jacob Hamblin, a Mormon missionary. These two points are so near each other that only one of them can be selected for the purpose above mentioned, and others must be found. We have been unable up to this time to obtain, either from Indians or white men, any information which will give us a clue to any other trail to the river.

At the headwaters of the Sevier, we are on the summit of a great watershed. The Sevier itself flows north and then westward into the lake of the same name. The Rio Virgen, heading near by, flows to the southwest into the Colorado, 60 or 70 miles below the Grand Canyon. The Kanab, also heading near by, runs directly south into the very heart of the Grand Canyon. The Paria, likewise heading near by, runs a little south of east and enters the river at the head of Marble Canyon. To the northeast from this point, other streams which run into the Colorado have their sources, until, 40 or 50 miles away, we reach the southern branches of the Dirty Devil River, the mouth of which stream is but a short distance below the junction of the Grand and Green.

The Paunsa'gunt Plateau terminates in a point, which is bounded by a line of beautiful pink cliffs. At the foot of this plateau, on the west, the Rio Virgen and Sevier River are dovetailed together, as their minute upper branches interlock. The upper surface of the plateau inclines to the northeast, so that its waters roll off into the Sevier; but from the foot of the cliffs, quite around the sharp angle of the plateau, for a dozen miles, we find numerous springs, whose waters unite to form the Kanab. A little farther to the northeast the springs gather into streams that feed the Paria. Here, by the upper springs of the Kanab, we make a camp, and from this point we are to radiate on a series of trips, southwest, south, and east.

Jacob Hamblin, who has been a missionary among the Indians for more than twenty years, has collected a number of Kai'vavits, with Chuar'-ruumpeak, their chief, and they are all camped with us. They assure us that we cannot reach the river, that we cannot make our way into the depths of the canyon, but promise to show us the springs and water pockets, which are very scarce in all this region, and to give us all the information in their power. Here we fit up a pack train, for our bedding and instruments and supplies are to be carried on the backs of mules and ponies.

September 5, 1870.–The several members of the party are engaged in general preparation for our trip down to the Grand Canyon.

Taking with me a white man and an Indian, I start on a climb to the summit of the Paunsa'gunt Plateau, which rises above us on the east. Our way for a mile or more is over a great peat bog, which trembles under our feet, and now and then a mule sinks through the broken turf and we are compelled to pull it out with ropes. Passing the bog, our way is up a gulch at the foot of the Pink Cliffs, which form the escarpment, or wall, of the great plateau. Soon we leave the gulch and climb a long ridge which winds around to the right toward the summit of the great table.

Two hours' riding, climbing, and clambering bring us near the top. We look below and see clouds drifting up from the south and rolling tumultuously toward the foot of the cliffs beneath us. Soon all the country below is covered with a sea of vapor–a billowy, raging, noiseless sea–and as the vapory flood still rolls up from the south, great waves dash against the foot of the cliffs and roll back; another tide comes in, is hurled back, and another and another, lashing the cliffs until the fog rises to the summit and covers us all. There is a heavy pine and fir forest above, beset with dead and fallen timber, and we make our way through the undergrowth to the east.

It rains. The clouds discharge their moisture in torrents, and we make for ourselves shelters of boughs, only to be soon abandoned, and we stand shivering by a great fire of pine logs and boughs, which the pelting storm half extinguishes.

One, two, three, four hours of the storm, and at last it partially abates. During this time our animals, which we have turned loose, have sought for themselves shelter under the trees, and two of them have wandered away beyond our sight. I go out to follow their tracks, and come near to the brink of a ledge of rocks, which, in the fog and mist, I suppose to be a little ridge, and I look for a way by which I can go down. Standing just here, there is a rift made in the fog below by some current or blast of wind, which reveals an almost bottomless abyss. I look from the brink of a great precipice of more than 2,000 feet; but through the mist the forms are half obscured and all reckoning of distance is lost, and it seems 10,000 feet, ten miles–any distance the imagination desires to make it.

Catching our animals, we return to the camp. We find that the little streams which come down from the plateau are greatly swollen, but at camp they have had no rain. The clouds which drifted up from the south, striking against the plateau, were lifted up into colder regions and discharged their moisture on the summit and against the sides of the plateau, but there was no rain in the valley below.

September 9.–We make a fair start this morning from the beautiful meadow at the head of the Kanab, cross the line of little hills at the headwaters of the Rio Virgen, and pass, to the south, a pretty valley. At ten o'clock we come to the brink of a great geographic bench–a line of cliffs. Behind us are cool springs, green meadows, and forest-clad slopes; below us, stretching to the south until the world is lost in blue haze, is a painted desert–not a desert plain, but a desert of rocks cut by deep gorges and relieved by towering cliffs and pinnacled rocks–naked rocks, brilliant in the sunlight.

By a difficult trail we make our way down the basaltic ledge, through which innumerable streams here gather into a little river running in a deep canyon. The river runs close to the foot of the cliffs on the right-hand side and the trail passes along to the right. At noon we rest and our animals feed on luxuriant grass.

Again we start and make slow progress along a stony way. At night we camp under an overarching cliff.

September 10.–Here the river turns to the west, and our way, properly, is to the south; but we wish to explore the Rio Virgen as far as possible. The Indians tell us that the canyon narrows gradually a few miles below and that it will be impossible to take our animals much farther down the river. Early in the morning I go down to examine the head of this narrow part. After breakfast, having concluded to explore the canyon for a i few miles on foot, we arrange that the main party shall climb the cliff and go around to a point 18 or 20 \ miles below, where, the Indians say, the animals can be taken down by the river, and three of us set out on, foot.

The Indian name of the canyon is Paru'nuweap, or Roaring Water Canyon. Between the little river and the foot of the walls is a dense growth of willows, vines, and wild rosebushes, and with great difficulty we make our way through this tangled mass. It is not a wide stream–only 20 or 30 feet across in most places; shallow, but very swift. After spending some hours in breaking our way through the mass of vegetation and climbing rocks here and there, it is determined to wade along the stream. In some places this is an easy task, but here and there we come to deep holes where we have to wade to our armpits. Soon we come to places so narrow that the river fills the entire channel and we wade perforce. In many places the bottom is a quicksand, into which we sink, and it is with great difficulty that we make progress. In some places the holes are so deep that we have to swim, and our little bundles of blankets and rations are fixed to a raft made of driftwood and pushed before us.

Now and then there is a little flood-plain, on which we can walk, and we cross and recross the stream and wade along the channel where the water is so swift as almost to carry us off our feet and we are in danger every moment of being swept down, until night comes on. Finding a little patch of flood-plain, on which there is a huge pile of driftwood and a clump of box-elders, and near by a mammoth stream bursting from the rocks, we soon have a huge fire. Our clothes are spread to dry; we make a cup of coffee, take out our bread and cheese and dried beef, and enjoy a hearty supper. We estimate that we have traveled eight miles to-day.

The canyon here is about 1,200 feet deep. It has been very narrow and winding all the way down to this point.

September 11.–Wading again this morning; sinking in the quicksand, swimming the deep waters, and making slow and painful progress where the waters are swift and the bed of the stream rocky.

The canyon is steadily becoming deeper and in many places very narrow–only 20 or 30 feet wide below, and in some places no wider, and even narrower, for hundreds of feet overhead. There are places where the river in sweeping by curves has cut far under the rocks, but still preserves its narrow channel, so that there is an overhanging wall on one side and an inclined wall on the other. In places a few hundred feet above, it becomes vertical again, and thus the view to the sky is entirely closed. Everywhere this deep passage is dark and gloomy and resounds with the noise of rapid waters. At noon we are in a canyon 2,500 feet deep, and we come to a fall where the walls are broken down and huge rocks beset the channel, on which we obtain a foothold to reach a level 200 feet below. Here the canyon is again wider, and we find a flood-plain along which we can walk, now on this, and now on that side of the stream. Gradually the canyon widens; steep rapids, cascades, and cataracts are found along the river, but we wade only when it is necessary to cross. We make progress with very great labor, having to climb over piles of broken rocks.

Late in the afternoon we come to a little clearing in the valley and see other signs of civilization and by sundown arrive at the Mormon town of Schunesburg; and here we meet the train, and feast on melons and grapes.

September 12.–Our course for the last two days, through Paru'nuweap Canyon, was directly to the west. Another stream comes down from the north and unites just here at Schunesburg with the main branch of the Rio Virgen. We determine to spend a day in the exploration of this stream. The Indians call the canyon through which it runs, Mukun'tu-weap, or Straight, Canyon. Entering this, we have to wade upstream; often the water fills the entire channel and, although we travel many miles, we find no flood-plain, talus, or broken piles of rock at the foot of the cliff. The walls have smooth, plain faces and are everywhere very regular and vertical for a thousand feet or more, where they seem to break back in shelving slopes to higher altitudes; and everywhere, as we go along, we find springs bursting out at the foot of the walls, and passing these the river above becomes steadily smaller. The great body of water which runs below bursts out from beneath this great bed of red sandstone; as we go up the canyon, it comes to be but a creek, and then a brook. On the western wall of the canyon stand some buttes, towers, and high pinnacled rocks. Going up the canyon, we gain glimpses of them, here and there. Last summer, after our trip through the canyons of the Colorado, on our way from the mouth of the Virgen to Salt Lake City, these were seen as conspicuous landmarks from a distance away to the southwest of 60 or 70 miles. These tower rocks are known as the Temples of the Virgen.

Having explored this canyon nearly to its head, we return to Schunesburg, arriving quite late at night.

Sitting in camp this evening, Chuar'ruumpeak, the chief of the Kai'vavits, who is one of our party, tells us there is a tradition among the tribes of this country that many years ago a great light was seen somewhere in this region by the Paru'shapats, who lived to the southwest, and that they supposed it to be a signal kindled to warn them of the approach of the Navajos, who lived beyond the Colorado River to the east. Then other signal fires were kindled on the Pine Valley Mountains, Santa Clara Mountains, and Uinkaret Mountains, so that all the tribes of northern Arizona, southern Utah, southern Nevada, and southern California were warned of the approaching danger; but when the Paru'shapats came nearer, they discovered that it was a fire on one of the great temples; and then they knew that the fire was not kindled by men, for no human being could scale the rocks. The Tu'muurrugwait'sigaip, or Rock Rovers, had kindled a fire to deceive the people. So, in the Indian language this is called Tu'muurruwait'sigaip Tuweap', or Rock Rovers' Land.

September 13.–We start very early this morning, for we have a long day's travel before us. Our way is across the Rio Virgen to the south. Coming to the bank of the stream here, we find a strange metamorphosis. The streams we have seen above, running in narrow channels, leaping and plunging over the rocks, raging and roaring in their course, are here united and spread in a thin sheet several hundred yards wide and only a few inches deep, but running over a bed of quicksand. Crossing the stream, our trail leads up a narrow canyon, not very deep, and then among the hills of golden, red, and purple shales and marls. Climbing out of the valley of the Rio Virgen, we pass through a forest of dwarf cedars and come out at the foot of the Vermilion Cliffs. All day we follow this Indian trail toward the east, and at night camp at a great spring, known to the Indians as Yellow Rock Spring, but to the Mormons as Pipe Spring; and near by there is a cabin in which some Mormon herders find shelter. Pipe Spring is a point just across the Utah line in Arizona, and we suppose it to be about 60 miles from the river. Here the Mormons design to build a fort another year, as an outpost for protection against the Indians. We now discharge a number of the Indians, but take two with us for the purpose of showing us the springs, for they are very scarce, very small, and not easily found. Half a dozen are not known in a district of country large enough to make as many good-sized counties in Illinois. There are no running streams, and these springs and water pockets are our sole dependence.

Starting, we leave behind a long line of cliffs, many hundred feet high, composed of orange and vermilion sandstones. I have named them "Vermilion Cliffs." When we are out a few miles, I look back and see the morning sun shining in splendor on their painted faces; the salient angles are on fire, and the retreating angles are buried in shade, and I gaze on them until my vision dreams and the cliffs appear a long bank of purple clouds piled from the horizon high into the heavens. At noon we pass along a ledge of chocolate cliffs, and, taking out our sandwiches, we make a dinner as we ride along.

Yesterday our Indians discussed for hours the route which we should take. There is one way, farther by 10 or 12 miles, with sure water; another, shorter, where water is found sometimes; their conclusion was that water would be found now; and this is the way we go, yet all day long we are anxious about it. To be out two days with only the water that can be carried in two small kegs is to have our animals suffer greatly. At five o'clock we come to the spot, and there is a huge water pocket containing several barrels. What a relief! Here we camp for the night.

September 15.–Up at daybreak, for it is a long day's march to the next water. They say we must "run very hard" to reach it by dark.

Our course is to the south. From Pipe Spring we can see a mountain, and I recognize it as the one seen last summer from a cliff overlooking the Grand Canyon; and I wish to reach the river just behind the mountain. There are Indians living in the group, of which it is the highest, whom I wish to visit on the way. These mountains are of volcanic origin, and we soon come to ground that is covered with fragments of lava. The way becomes very difficult. We have to cross deep ravines, the heads of canyons that run into the Grand Canyon. It is curious now to observe the knowledge of our Indians. There is not a trail but what they know; every gulch and every rock seems familiar. I have prided myself on being able to grasp and retain in my mind the topography of a country; but these Indians put me to shame. My knowledge is only general, embracing the more important features of a region that remains as a map engraved on my mind; but theirs is particular. They know every rock and every ledge, every gulch and canyon, and just where to wind among these to find a pass; and their knowledge is unerring. They cannot describe a country to you, but they can tell you all the particulars of a route.

I have but one pony for the two, and they were to ride "turn about"; but Chuar'ruumpeak, the chief, rides, and Shuts, the one-eyed, barelegged, merry-faced pigmy, walks, and points the way with a slender cane; then leaps and bounds by the shortest way, and sits down on a rock and waits demurely until we come, always meeting us with a jest, his face a rich mine of sunny smiles.

At dusk we reach the water pocket. It is in a deep gorge on the flank of this great mountain. During the rainy season the water rolls down the mountain side, plunging over precipices, and excavates a deep basin in the solid rock below. This basin, hidden from the sun, holds water the year round.

September 16.–This morning, while the men are packing the animals, I climb a little mountain near camp, to obtain a view of the country. It is a huge pile of volcanic scoria, loose and light as cinders from a forge, which give way under my feet, and I climb with great labor; but, reaching the summit and looking to the southeast, I see once more the labyrinth of deep gorges that flank the Grand Canyon; in the multitude, I cannot determine whether it is itself in view or not. The memories of grand and awful months spent in their deep, gloomy solitudes come up, and I live that life over again for a time.

I supposed, before starting, that I could get a good view of the great mountain from this point; but it is like climbing a chair to look at a castle. I wish to discover some way by which it can be ascended, as it is my intention to go to the summit before I return to the settlements. There is a cliff near the summit and I do not see any way yet. Now down I go, sliding on the cinders, making them rattle and clang.

The Indians say we are to have a short ride to-day and that we shall reach an Indian village, situated by a good spring. Our way is across the spurs that put out from the great mountain as we pass it to the left.

Up and down we go across deep ravines, and the fragments of lava clank under our horses' feet; now among cedars, now among pines, and now across mountain-side glades. At one o'clock we descend into a lovely valley, with a carpet of waving grass; sometimes there is a little water in the upper end of it, and during some seasons the Indians we wish to find are encamped here. Chuar'ruumpeak rides on to find them, and to say we are friends, otherwise they would run away or propose to fight us, should we come without notice. Soon we see Chuar'ruumpeak riding at full speed and hear him shouting at the top of his voice, and away in the distance are two Indians scampering up the mountain side. One stops; the other still goes on and is soon lost to view. We ride up and find Chuar'ruumpeak talking with the one who had stopped. It is one of the ladies resident in these mountain glades; she is evidently paying taxes, Godiva-like. She tells us that her people are at the spring; that it is only two hours' ride; that her good master has gone on to tell them we are coming; and that she is harvesting seeds.

We sit down and eat our luncheon and share our biscuits with the woman of the mountains; then on we go over a divide between two rounded peaks. I send the party on to the village and climb the peak on the left, riding my horse to the upper limit of trees and then tugging up afoot. From this point I can see the Grand Canyon, and I know where I am. I can see the Indian village, too, in a grassy valley, embosomed in the mountains, the smoke curling up from their fires; my men are turning out their horses and a group of natives stand around. Down the mountain I go and reach camp at sunset. After supper we put some cedar boughs on the fire; the dusky villagers sit around, and we have a smoke and a talk. I explain the object of my visit, and assure them of my friendly intentions. Then I ask them about a way down into the canyon. They tell me that years ago a way was discovered by which parties could go down, but that no one has attempted it for a long time; that it is a very difficult and very dangerous undertaking to reach the "Big Water." Then I inquire about the Shi'vwits, a tribe that lives about the springs on the mountain sides and canyon cliffs to the southwest. They say that their village is now about 30 miles away, and promise to send a messenger for them to-morrow morning.

Having finished our business for the evening, I ask if there is a tugwi'nagunt in camp; that is, if there is any one present who is skilled in relating their mythology. Chuar'ruumpeak says Tomor'rountikai, the chief of these Indians, is a very noted man for his skill in this matter; but they both object, by saying that the season for tugwi'nai has not yet arrived. But I had anticipated this, and soon some members of the party come with pipes and tobacco, a large kettle of coffee, and a tray of biscuits, and, after sundry ceremonies of pipe lighting and smoking, we all feast, and, warmed up by this, to them, unusually good living, it is decided that the night shall be spent in relating mythology. I ask Tomor'rountikai to tell us about the So'kus Wai'unats, or One-Two Boys, and to this he agrees.

The long winter evenings of an Indian camp are usually devoted to the relation of mythologic stories, which purport to give a history of an ancient race of animal gods. The stories are usually told by some old man, assisted by others of the party, who take secondary parts, while the members of the tribe gather about and make comments or receive impressions from the morals which are enforced by the story-teller, or, more properly, story-tellers; for the exercise partakes somewhat of the nature of a theatrical performance.

THE SO'KUS WAI'UNATS.

Tumpwinai'rogwinump, He Who Had A Stone Shirt, killed Sikor', the Crane, and stole his wife, and seeing that she had a child and thinking it would be an incumbrance to them on their travels, he ordered her to kill it. But the mother, loving the babe, hid it under her dress and carried it away to its grandmother. And Stone Shirt carried his captured bride to his own land.

In a few years the child grew to be a fine lad, under the care of his grandmother, and was her companion wherever she went.

One day they were digging flag roots on the margin of the river and putting them in a heap on the bank. When they had been at work a little while, the boy perceived that the roots came up with greater ease than was customary and he asked the old woman the cause of this, but she did not know; and, as they continued their work, still the reeds came up with less effort, at which their wonder increased, until the grandmother said,

"Surely, some strange thing is about to transpire."

Then the boy went to the heap where they had been placing the roots, and found that some one had taken them away, and he ran back, exclaiming,

"Grandmother, did you take the roots away?"

And she answered,

"No, my child; perhaps some ghost has taken them off; let us dig no more; come away."

But the boy was not satisfied, as he greatly desired to know what all this meant; so he searched about for a time, and at length found a man sitting under a tree, and taunted him with being a thief, and threw mud and stones at him until he broke the stranger's leg. The man answered not the boy nor resented the injuries he received, but remained silent and sorrowful; and when his leg was broken he tied it up in sticks and bathed it in the river and sat down again under the tree and beckoned the boy to approach. When the lad came near, the stranger told him he had something of great importance to reveal.

"My son," said he, "did that old woman ever tell you about your father and mother?"

"No," answered the boy; "I have never heard of them."

"My son, do you see these bones scattered on the ground? Whose bones are these?"

"How should I know?" answered the boy. "It may be that some elk or deer has been killed here."

"No," said the old man.

"Perhaps they are the bones of a bear"; but the old man shook his head.

So the boy mentioned many other animals, but the stranger still shook his head, and finally said,

"These are the bones of your father; Stone Shirt killed him and left him to rot here on the ground like a wolf."

And the boy was filled with indignation against the slayer of his father.

Then the stranger asked,

"Is your mother in yonder lodge?"

"No," the boy replied.

"Does your mother live on the banks of this river?"

"I don't know my mother; I have never seen her; she is dead," answered the boy.

"My son," replied the stranger, "Stone Shirt, who killed your father, stole your mother and took her away to the shore of a distant lake, and there she is his wife to-day."

And the boy wept bitterly and, while the tears filled his eyes so that he could not see, the stranger disappeared. Then the boy was filled with wonder at what he had seen and heard, and malice grew in his heart against his father's enemy. He returned to the old woman and said,

"Grandmother, why have you lied to me about my father and mother?"

But she answered not, for she knew that a ghost had told all to the boy. And the boy fell upon the ground weeping and sobbing, until he fell into a deep sleep, when strange things were told him.

His slumber continued three days and three nights and when he awoke he said to his grandmother:

"I am going away to enlist all nations in my fight."

And straightway he departed.

(Here the boy's travels are related with many circumstances concerning the way he was received by the people, all given in a series of conversations, very lengthy; so they will be omitted.)

Finally he returned in advance of the people whom he had enlisted, bringing with him Shinau'av, the Wolf, and Togo'av, the Rattlesnake. When the three had eaten food, the boy said to the old woman:

"Grandmother, cut me in two!"

But she demurred, saying she did not wish to kill one whom she loved so dearly.

"Cut me in two!" demanded the boy; and he gave her a stone ax, which he had brought from a distant country, and with a manner of great authority he again commanded her to cut him in two. So she stood before him and severed him in twain and fled in terror. And lo! each part took the form of an entire man, and the one beautiful boy appeared as two, and they were so much alike no one could tell them apart.

When the people or natives whom the boy had enlisted came pouring into the camp, Shinau'av and Togo'av were engaged in telling them of the wonderful thing that had happened to the boy, and that now there were two; and they all held it to be an augury of a successful expedition to the land of Stone Shirt. And they started on their journey.

Now the boy had been told in the dream of his three days'

slumber, of a magical cup, and he had brought it home with him from his journey among the nations, and the So'kus Wai'unats carried it between them, filled with water. Shinau'av walked on their right and Togo'av on their left, and the nations followed in the order in which they had been enlisted. There was a vast number of them, so that when they were stretched out in line it was one day's journey from the front to the rear of the column.

When they had journeyed two days and were far out on the desert, all the people thirsted, for they found no water, and they fell down upon the sand groaning and murmuring that they had been deceived, and they cursed the One-Two.

But the So'kus Wai'unats had been told in the wonderful dream of the suffering which would be endured, and that the water which they carried in the cup was to be used only in dire necessity; and the brothers said to each other:

"Now the time has come for us to drink the water."

And when one had quaffed of the magical bowl, he found it still full; and he gave it to the other to drink, and still it was full; and the One-Two gave it to the people, and one after another did they all drink, and still the cup was full to the brim.

But Shinau'av was dead, and all the people mourned, for he was a great man. The brothers held the cup over him and sprinkled him with water, when he arose and said:

"Why do you disturb me? I did have a vision of mountain brooks and meadows, of cane where honey dew was plenty."

They gave him the cup and he drank also; but when he had finished there was none left. Refreshed and rejoicing, they proceeded on their journey.

The next day, being without food, they were hungry, and all were about to perish; and again they murmured at the brothers and cursed them. But the So'kus Wai'unats saw in the distance an antelope, standing on an eminence in the plain, in bold relief against the sky; and Shinau'av knew it was the wonderful antelope with many eyes which Stone Shirt kept for his watchman; and he proposed to go and kill it, but Togo'av demurred and said:

"It were better that I should go, for he will see you and run away."

But the So'kus Wai'unats told Shinau'av to go; and he started in a direction away to the left of where the antelope was standing, that he might make a long detour about some hills and come upon him from the other side.

Togo'av went a little way from camp and called to the brothers:

"Do you see me!"

They answered they did not.

"Hunt for me."

While they were hunting for him, the Rattlesnake said:

"I can see you; you are doing so and so," telling them what they were doing; but they could not find him.

Then the Rattlesnake came forth declaring:

"Now you know that when I so desire I can see others and I cannot be seen. Shinau'av cannot kill that antelope, for he has many eyes, and is the wonderful watchman of Stone Shirt; but I can kill him, for I can go where he is and he cannot see me."

So the brothers were convinced and permitted him to go; and Togo'av went and killed the antelope. When Shinau'av saw it fall, he was very angry, for he was extremely proud of his fame as a hunter and anxious to have the honor of killing this famous antelope, and he ran up with the intention of killing Togo'av; but when he drew near and saw the antelope was fat and would make a rich feast for the people, his anger was appeased.

"What matters it," said he, "who kills the game, when we can all eat it?"

So all the people were fed in abundance and they proceeded on their journey.

The next day the people again suffered for water, and the magical cup was empty; but the So'kus Wai'unats, having been told in their dream what to do, transformed themselves into doves and flew away to a lake, on the margin of which was the home of Stone Shirt.

Coming near to the shore, they saw two maidens bathing in the water; and the birds stood and looked, for the maidens were very beautiful. Then they flew into some bushes near by, to have a nearer view, and were caught in a snare which the girls had placed for intrusive birds.

The beautiful maidens came up and, taking the birds out of the snare, admired them very much, for they had never seen such birds before. They carried them to their father, Stone Shirt, who said:

"My daughters, I very much fear these are spies from my enemies, for such birds do not live in our land."

He was about to throw them into the fire, when the maidens besought him, with tears, that he would not destroy their beautiful birds; but he yielded to their entreaties with much misgiving. Then they took the birds to the shore of the lake and set them free.

When the birds were at liberty once more they flew around among the bushes until they found the magical cup which they had lost, and taking it up they carried it out into the middle of the lake and settled down upon the water, and the maidens supposed they were drowned.

The birds, when they had filled their cup, rose again and went back to the people in the desert, where they arrived just at the right time to save them with the cup of water, from which each drank;

and yet it was full until the last was satisfied, and then not a drop remained.

The brothers reported that they had seen Stone Shirt and his daughters.

The next day they came near to the home of the enemy, and the brothers, in proper person, went out to reconnoiter. Seeing a woman gleaning seeds, they drew near, and knew it was their mother, whom Stone Shirt had stolen from Sikor', the Crane. They told her they were her sons, but she denied it and said she had never had but one son; but the boys related to her their history, with the origin of the two from one, and she was convinced. She tried to dissuade them from making war upon Stone Shirt, and told them that no arrow could possibly penetrate his armor, and that he was a great warrior and had no other delight than in killing his enemies, and that his daughters also were furnished with magical bows and arrows, which they could shoot so fast that the arrows would fill the air like a cloud, and that it was not necessary for them to take aim, for their missiles went where they willed; they thought the arrows to the hearts of their enemies; and thus the maidens could kill the whole of the people before a common arrow could be shot by a common person. But the boys told her what the spirit had said in the long dream and that it had promised that Stone Shirt should be killed. They told her to go down to the lake at dawn, so as not to be endangered by the battle.

During the night the So'kus Wai'unats transformed themselves into mice and proceeded to the home of Stone Shirt and found the magical bows and arrows that belonged to the maidens, and with their sharp teeth they cut the sinew on the backs of the bows and nibbled the bow strings, so that they were worthless. Togo'av hid himself under a rock near by.

When dawn came into the sky, Tumpwinai'ro-gwinump, the Stone Shirt man, arose and walked out of his tent, exulting in his strength and security, and sat down upon the rock under which Togo'av was hiding; and he, seeing his opportunity, sank his fangs into the flesh of the hero. Stone Shirt sprang high into the air and called to his daughters that they were betrayed and that the enemy was near; and they seized their magical bows and their quivers filled with magical arrows and hurried to his defense. At the same time, all the nations who were surrounding the camp rushed down to battle. But the beautiful maidens, finding their weapons were destroyed, waved back their enemies, as if they would parley; and standing for a few moments over the body of their slain father, sang the death song and danced the death dance, whirling in giddy circles about the dead hero and wailing with despair, until they sank down and expired.

The conquerors buried the maidens by the shores of the lake; but Tumpwinai'rogwinump was left to rot and his bones to bleach on the sands, as he had left Sikor'.

There is this proverb among the Utes: "Do not murmur when you suffer in doing what the spirits have commanded, for a cup of water is provided"; and another: "What matters it who kills the game, when we can all eat of it?"

It is long after midnight when the performance is ended. The story itself is interesting, though I had heard it many times before; but never, perhaps, under circumstances more effective. Stretched beneath tall, somber pines; a great camp fire; by the fire, men, old, wrinkled, and ugly; deformed, blear-eyed, wry-faced women; lithe, stately young men; pretty but simpering maidens, naked children, all intently listening, or laughing and talking by turns, their strange faces and dusky forms lit up with the glare of the pine-knot fire. All the circumstances conspired to make it a scene strange and weird. One old man, the sorcerer or medicine man of the tribe, peculiarly impressed me. Now and then he would interrupt the play for the purpose of correcting the speakers or impressing the moral of the story with a strange dignity and impressiveness that seemed to pass to the very border of the ludicrous; yet at no time did it make me smile.

The story is finished, but there is yet time for an hour or two of sleep. I take Chuar'ruumpeak to one side for a talk. The three men who left us in the canyon last year found their way up the lateral gorge, by which they went into the Shi'wits Mountains, lying west of us, where they met with the Indians and camped with them one or two nights and were finally killed. I am anxious to learn the circumstances, and as the people of the tribe who committed the deed live but a little way from these people and are intimate with them, I ask Chuar'ruumpeak to make inquiry for me. Then we go to bed.

September 17.–Early this morning the Indians come up to our camp.

They have concluded to send out a young man after the Shi'vwits. The runner fixes his moccasins, puts some food in a sack and water in a little wickerwork jug, straps them on his back, and starts at a good round pace.

We have concluded to go down the canyon, hoping to meet the Shi'vwits on our return. Soon we are ready to start, leaving the camp and pack animals in charge of the two Indians who came with us. As we move out our new guide comes up, a blear-eyed, weazen-faced, quiet old man, with his bow and arrows in one hand and a small cane in the other. These Indians all carry canes with a crooked handle, they say to kill rattlesnakes and to pull rabbits from their holes. The valley is high up in the mountain and we descend from it by a rocky, precipitous trail, down, down, down for two long, weary hours, leading our ponies and stumbling over the rocks. At last we are at the foot of the mountain, standing on a little knoll, from which we can look into a canyon below.

Into this we descend, and then we follow it for miles, clambering down and still down. Often we cross beds of lava, that have been poured into the canyon by lateral channels, and these angular fragments of basalt make the way very rough for the animals.

About two o'clock the guide halts us with his wand, and, springing over the rocks, he is lost in a gulch. In a few minutes he returns, and tells us there is a little water below in a pocket. It is vile and our ponies refuse to drink it. We pass on, still descending. A mile or two from the water basin we come to a precipice more than 1,000 feet to the bottom. There is a canyon running at a greater depth and at right angles to this, into which this enters by the precipice; and this second canyon is a lateral one to the greater one, in the bottom of which we are to find the river. Searching about, we find a way by which we can descend along the shelves and steps and piles of broken rocks.

We start, leading our ponies; a wall upon our left; unknown depths on our right. At places our way is along shelves so narrow or so sloping that I ache with fear lest a pony should make a misstep and knock a man over the cliffs with him. Now and then we start the loose rocks under our feet, and over the cliffs they go, thundering down, down, the echoes rolling through distant canyons. At last we pass along a level shelf for some distance, then we turn to the right and zigzag down a steep slope to the bottom. Now we pass along this lower canyon for two or three miles, to where it terminates in the Grand Canyon, as the other ended in this, only the river is 1,800 feet below us, and it seems at this distance to be but a creek. Our withered guide, the human pickle, seats himself on a rock and seems wonderfully amused at our discomfiture, for we can see no way by which to descend to the river. After some minutes he quietly rises and, beckoning us to follow, points out a narrow sloping shelf on the right, and this is to be our way. It leads along the cliff for half a mile to a wider bench beyond, which, he says, is broken down on the other side in a great slide, and there we can get to the river. So we start out on the shelf; it is so steep we can hardly stand on it, and to fall or slip is to go–don't look to see!

It is soon manifest that we cannot get the ponies along the ledge. The storms have washed it down since our guide was here last, years ago. One of the ponies has gone so far that we cannot turn him back until we find a wider place, but at last we get him off. With part of the men, I take the horses back to the place where there are a few bushes growing and turn them loose; in the meantime the other men are looking for some way by which we can get down to the river. When I return, one, Captain Bishop, has found a way and gone down. We pack bread, coffee, sugar, and two or three blankets among us, and set out. It is now nearly dark, and we cannot find the way by which the captain went, and an hour is spent in fruitless search. Two of the men go away around an amphitheater, more than a fourth of a mile, and start down a broken chasm that faces us who are behind. These walls, that are vertical, or nearly so, are often cut by chasms, where the showers run down, and the top of these chasms will be back a distance from the face of the wall, and the bed of the chasm will slope down, with here and there a fall. At other places huge rocks have fallen and block the way. Down such a one the two men start. There is a curious plant growing out from the crevices of the rock. A dozen stems will start from one root and grow to the length of eight or ten feet and not throw out a branch or twig, but these stems are thickly covered with leaves. Now and then the two men come to a bunch of dead stems and make a fire to mark for us their way and progress.

In the meantime we find such a gulch and start down, but soon come to the "jumping-off place," where we can throw a stone and faintly hear it strike, away below. We fear that we shall have to stay here, clinging to the rocks until daylight. Our little Indian gathers a few dry stems, ties them into a bundle, lights one end, and holds it up. The others do the same, and with these torches we find a way out of trouble. Helping each other, holding torches for each other, one clinging to another's hand until we can get footing, then supporting the other on his shoulders, thus we make our passage into the depths of the canyon.

And now Captain Bishop has kindled a huge fire of driftwood on the bank of the river. This and the fires in the gulch opposite and our own flaming torches light up little patches that make more manifest the awful darkness below. Still, on we go for an hour or two, and at last we see Captain Bishop coming up the gulch with a huge torchlight on his shoulders. He looks like a fiend, waving brands and lighting the fires of hell, and the men in the opposite gulch are imps, lighting delusive fires in inaccessible crevices, over yawning chasms; our own little Indian is surely the king of wizards, so I think, as I stop for a few moments on a rock to rest. At last we meet Captain Bishop, with his flaming torch, and as he has learned the way he soon pilots us to the side of the great Colorado. We are athirst and hungry, almost to starvation. Here we lie down on the rocks and drink, just a mouthful or so, as we dare; then we make a cup of coffee, and spreading our blankets on a sand beach the roaring Colorado lulls us to sleep.

September 18.–We are in the Grand Canyon, by the side of the Colorado, more than 6,000 feet below our camp on the mountain side, which is 18 miles away; but the miles of horizontal distance represent but a small part of the day's labor before us. It is the mile of altitude we must gain that makes it a Herculean task. We are up early; a little bread and coffee, and we look about us. Our conclusion is that we can make this a depot of supplies, should it be necessary; that we can pack our rations to the point where we left our animals last night, and that we can employ Indians to bring them down to the water's edge.

On a broad shelf we find the ruins of an old stone house, the walls of which are broken down, and we can see where the ancient people who lived here–a race more highly civilized than the present–had made a garden and used a great spring that comes out of the rocks for irrigation. On some rocks near by we discover some curious etchings. Still searching about, we find an obscure trail up the canyon wall, marked here and there by steps which have been built in the loose rock, elsewhere hewn stairways, and we find a much easier way to go up than that by which we came down in the darkness last night. Coming to the top of the wall, we catch our horses and start. Up the canyon our jaded ponies toil and we reach the second cliff; up this we go, by easy stages, leading the animals. Now we reach the offensive water pocket; our ponies have had no water for thirty hours, and are eager even for this foul fluid. We carefully strain a kettleful for ourselves, then divide what is left between them–two or three gallons for each; but it does not satisfy them, and they rage around, refusing to eat the scanty grass. We boil our kettle of water, and skim it; straining, boiling, and skimming make it a little better, for it was full of loathsome, wriggling larvae, with huge black heads. But plenty of coffee takes away the bad smell, and so modifies the taste that most of us can drink, though our little Indian seems to prefer the original mixture. We reach camp about sunset, and are glad to rest.

September 19.–We are tired and sore, and must rest a day with our Indian neighbors. During the inclement season they live in shelters made of boughs or the bark of the cedar, which they strip off in long shreds. In this climate, most of the year is dry and warm, and during such time they do not care for shelter. Clearing a small, circular space of ground, they bank it around with brush and sand, and wallow in it during the day and huddle together in a heap at night–men, women, and children; buckskin, rags, and sand. They wear very little clothing, not needing much in this lovely climate.

Altogether, these Indians are more nearly in their primitive condition than any others on the continent with whom I am acquainted. They have never received anything from the government and are too poor to tempt the trader, and their country is so nearly inaccessible that the white man never visits them. The sunny mountain side is covered with: wild fruits, nuts, and native grains, upon which they subsist. The oose, the fruit of the yucca, or Spanish bayonet, is rich, and not unlike the pawpaw of the valley of the Ohio. They eat it raw and also roast it in the ashes. They gather the fruits of a cactus plant, which are rich and luscious, and eat them as grapes or express the juice from them, making the dry pulp into cakes and saving them for winter and drinking the wine about their camp fires until the midnight is merry with their revelries.

They gather the seeds of many plants, as sunflowers, golden-rod, and grasses. For this purpose they have large conical baskets, which hold two or more bushels. The women carry them on their backs, suspended from their foreheads by broad straps, and with a smaller one in the left hand and a willow-woven fan in the right they walk among the grasses and sweep the seed into the smaller basket, which is emptied now and then into the larger, until it is full of seeds and chaff; then they winnow out the chaff and roast the seeds. They roast these curiously; they put seeds and a quantity of red-hot coals into a willow tray and, by rapidly and dexterously shaking and tossing them, keep the coals aglow and the seeds and tray from burning. So skilled are the crones in this work they roll the seeds to one side of the tray as they are roasted and the coals to the other as if by magic.

Then they grind the seeds into a fine flour and make it into cakes and mush. It is a merry sight, sometimes, to see the women grinding at the mill. For a mill, they use a large flat rock, lying on the ground, and another small cylindrical one in their hands. They sit prone on the ground, hold the large flat rock between the feet and legs, then fill their laps with seeds, making a hopper to the mill with their dusky legs, and grind by pushing the seeds across the larger rock, where they drop into a tray. I have seen a group of women grinding together, keeping time to a chant, or gossiping and chatting, while the younger lassies would jest and chatter and make the pine woods merry with their laughter.

Mothers carry their babes curiously in baskets. They make a wicker board by plaiting willows and sew a buckskin cloth to either edge, and this is fulled in the middle so as to form a sack closed at the bottom. At the top they make a wicker shade, like "my grandmother's sunbonnet," and wrapping the little one in a wild-cat robe, place it in the basket, and this they carry on their backs, strapped over the forehead, and the little brown midgets are ever peering over their mothers' shoulders. In camp, they stand the basket against the trunk of a tree or hang it to a limb.

There is little game in the country, yet they get a mountain sheep now and then or a deer, with their arrows, for they are not yet supplied with guns. They get many rabbits, sometimes with arrows, sometimes with nets. They make a net of twine, made of the fibers of a native flax. Sometimes this is made a hundred yards in length, and is placed in a half-circular position, with wings of sage brush. Then they have a circle hunt, and drive great numbers of rabbits into the snare, where they are shot with arrows. Most of their bows are made of cedar, but the best are made of the horns of mountain sheep. These are soaked in water until quite soft, cut into long thin strips, and glued together; they are then quite elastic. During the autumn, grasshoppers are very abundant, can be gathered by the bushel. At such a time, they dig a hole in the sand, heat stones in a fire near by, put some hot stones in the bottom of the hole, put on a layer of grasshoppers, then a layer of hot stones, and continue this, until they put bushels on to roast. There they are.

When cold weather sets in, these insects are numbed and left until cool, when they are taken out, thoroughly dried, and ground into meal. Grasshopper gruel or grasshopper cake is a great treat.

Their lore consists of a mass of traditions, or mythology. It is very difficult to induce them to tell it to white men; but the old Spanish priests, in the days of the conquest of New Mexico, spread among the Indians of this country many Bible stories, which the Indians are usually willing to tell. It is not always easy to recognize them; the Indian mind is a strange receptacle for such stories and they are apt to sprout new limbs. Maybe much of their added quaint-ness is due to the way in which they were told by the "fathers." But in a confidential way, while alone, or when admitted to their camp fire on a winter night, one may hear the stories of their mythology. I believe that the greatest mark of friendship or confidence that an Indian can give is to tell you his religion. After one has so talked with me I should ever trust him; and I feel on very good terms with these Indians since our experience of the other night.

A knowledge of the watering places and of the trails and passes is considered of great importance and is necessary to give standing to a chief.

This evening, the Shi'vwits, for whom we have sent, come in, and after supper we hold a long council. A blazing fire is built, and around this we sit–the Indians living here, the Shi'vwits, Jacob Hamblin, and myself.

This man, Hamblin, speaks their language well and has a great influence over all the Indians in the region round about. He is a silent, reserved man, and when he speaks it is in a slow, quiet way that inspires great awe. His talk is so low that they must listen attentively to hear, and they sit around him in deathlike silence. When he finishes a measured sentence the chief repeats it and they all give a solemn grunt. But, first, I fill my pipe, light it, and take a few whiffs, then pass it to Hamblin; he smokes, and gives it to the man next, and so it goes around. When it has passed the chief, he takes out his own pipe, fills and lights it, and passes it around after mine. I can smoke my own pipe in turn, but when the Indian pipe comes around, I am nonplused. It has a large stem, which has at some time been broken, and now there is a buckskin rag wound around it and tied with sinew, so that the end of the stem is a huge mouthful, exceedingly repulsive. To gain time, I refill it, then engage in very earnest conversation, and, all unawares, I pass it to my neighbor unlighted.

I tell the Indians that I wish to spend some months in their country during the coming year and that I would like them to treat me as a friend. I do not wish to trade; do not want their lands. Heretofore I have found it very difficult to make the natives understand my object, but the gravity of the Mormon missionary helps me much. I tell them that all the great and good white men are anxious to know very many things, that they spend much time in learning, and that the greatest man is he who knows the most; that the white men want to know all about the mountains and the valleys, the rivers and the canyons, the beasts and birds and snakes. Then I tell them of many Indian tribes, and where they live; of the European nations; of the Chinese, of Africans, and all the strange things about them that come to my mind. I tell them of the ocean, of great rivers and high mountains, of strange beasts and birds. At last I tell them I wish to learn about their canyons and mountains, and about themselves, to tell other men at home; and that I want to take pictures of everything and show them to my friends. All this occupies much time, and the matter and manner make a deep impression.

Then their chief replies: "Your talk is good, and we believe what you say. We believe in Jacob, and look upon you as a father. When you are hungry, you may have our game. You may gather our sweet fruits. We will give you food when you come to our land. We will show you the springs and you may drink; the water is good. We will be friends and when you come we will be glad. We will tell the Indians who live on the other side of the great river that we have seen Ka'purats, and that he is the Indians' friend. We will tell them he is Jacob's friend. We are very poor. Look at our women and children; they are naked. We have no horses; we climb the rocks and our feet are sore. We live among rocks and they yield little food and many thorns. When the cold moons come, our children are hungry. We have not much to give; you must not think us mean.

You are wise; we have heard you tell strange things. We are ignorant. Last year we killed three white men. Bad men said they were our enemies. They told great lies. We thought them true. Wo were mad; it made us big fools. We are very sorry. Do not think of them; it is done; let us be friends. We are ignorant–like little children in understanding compared with you. When we do wrong, do not you get mad and be like children too.

"When white men kill our people, we kill them. Then they kill more of us. It is not good. We hear that the white men are a great number. When they stop killing us, there will be no Indian left to bury the dead. We love our country; we know not other lands. We hear that other lands are better; we do not know. The pines sing and we are glad. Our children play in the warm sand; we hear them sing and are glad. The seeds ripen and we have to eat and we are glad. We do not want their good lands; we want our rocks and the great mountains where our fathers lived. We are very poor; we are very ignorant; but we are very honest. You have horses and many things. You are very wise; you have a good heart. We will be friends. Nothing more have I to say."

Ka'purats is the name by which I am known among the Utes and Shoshones, meaning "arm off." There was much more repetition than I have given, and much emphasis. After this a few presents were given, we shook hands, and the council broke up.

Mr. Hamblin fell into conversation with one of the men and held him until the others had left, and then learned more of the particulars of the death of the three men. They came upon the Indian village almost starved and exhausted with fatigue. They were supplied with food and put on their way to the settlements. Shortly after they had left, an Indian from the east side of the Colorado arrived at their village and told them about a number of miners having killed a squaw in drunken brawl, and no doubt these were the men; no person had ever come down the canyon; that was impossible; they were trying to hide their guilt. In this way he worked them into a great rage. They followed, surrounded the men in ambush, and filled them full of arrows.

That night I slept in peace, although these murderers of my men, and their friends, the Uinkarets, were sleeping not 500 yards away. While we were gone to the canyon, the pack train and supplies, enough to make an Indian rich beyond his wildest dreams, were all left in their charge, and were all safe; not even a lump of sugar was pilfered by the children.

September 20.–For several days we have been discussing the relative merits of several names for these mountains. The Indians call them Uinkarets, the region of pines, and we adopt the name. The great mountain we call Mount Trumbull, in honor of the senator. To-day the train starts back to the canyon water pocket, while Captain Bishop and I climb Mount Trumbull. On our way we pass the point that was the last opening to the volcano.

It seems but a few years since the last flood of fire swept the valley. Between two rough, conical hills it poured, and ran down the valley to the foot of a mountain standing almost at the lower end, then parted, and ran on either side of the mountain. This last overflow is very plainly marked; there is soil, with trees and grass, to the very edge of it, on a more ancient bed. The flood was, everywhere on its border, from 10 to 20 feet in height, terminating abruptly and looking like a wall from below. On cooling, it shattered into fragments, but these are still in place and the outlines of streams and waves can be seen. So little time has elapsed since it ran down that the elements have not weathered a soil, and there is scarcely any vegetation on it, but here and there a lichen is found. And yet, so long ago was it poured from the depths, that where ashes and cinders have collected in a few places, some huge cedars have grown. Near the crater the frozen waves of black basalt are rent with deep fissures, transverse to the direction, of the flow. Then we ride through a cedar forest up a long ascent, until we come to cliffs of columnar basalt. Here we tie our horses and prepare for a climb among the columns. Through crevices we work, till at last we are on the mountain, a thousand acres of pine laud spread out before us, gently rising to the other edge. There are two peaks on the mountain. We walk two miles to the foot of the one looking to be the highest, then a long, hard climb to its summit. What a view is before us! A vision of glory! Peaks of lava all around below us. The

Vermilion Cliffs to the north, with their splendor of colors; the Pine Valley Mountains to the northwest, clothed in mellow, perspective haze; unnamed mountains to the southwest, towering over canyons bottomless to my peering gaze, like chasms to nadir hell; and away beyond, the San Francisco Mountains, lifting their black heads into the heavens. We find our way down the mountain, reaching the trail made by the pack train just at dusk, and follow it through the dark until we see the camp fire–a welcome sight.

Two days more, and we are at Pipe Spring; one day, and we are at Kanab. Eight miles above the town is a canyon, on either side of which is a group of lakes. Four of these are in caves where the sun never shines. By the side of one of these I sit, at my feet the crystal waters, of which I may drink at will.

CHAPTER XIII. OVER THE RIVER.

It is our intention to explore a route from Kanab to the Colorado River at the mouth of the Paria, and, if successful in this undertaking, to cross the river and proceed to Tusayan, and ultimately to Santa Fe, New Mexico. We propose to build a flatboat for the purpose of ferrying over the river, and have had the lumber necessary for that purpose hauled from St. George to Kanab. From here to the mouth of the Paria it must be packed on the backs of mules; Captain Bishop and Mr. Graves are to take charge of this work, while with Mr. Hamblin I explore the Kaibab Plateau.

September 24–To-day we are ready for the start. The mules are packed and away goes our train of lumber, rations, and camping equipage. The Indian trail is at the foot of the Vermilion Cliffs. Pushing on to the east with Mr. Hamblin for a couple of hours in the early morning, we reach the mouth of a dry canyon, which comes down through the cliffs. Instead of a narrow canyon we find an open valley from one fourth to one half a mile in width. On rare occasions a stream flows down this valley, but now sand dunes stretch across it. On either side there is a wall of vertical rock of orange sandstone, and here and there at the foot of the wall are found springs that afford sweet water.

We push our way far up the valley to the foot of the Gray Cliffs, and by a long detour find our way to the summit. Here again we find that wonderful scenery of naked white rocks carved into great round bosses and domes. Looking off to the north we can see vermilion and pink cliffs, crowned with forests, while below us to the south stretch the dunes and red-lands of the Vermilion Cliff region, and far away we can see the opposite wall of the Grand Canyon. In the middle of the afternoon we descend into the canyon valley and hurriedly ride, down to the mouth of the canyon, then follow the trail of the pack train, for we are to camp with the party to-night. We find it at the Navajo Well. As we approach in the darkness the camp fire is a cheerful sight. The Navajo Well is a pool in the sand, the sands themselves lying in a basin, with naked, smooth rocks all about on which the rains are caught and by which the sand in the basin is filled with water, and by digging into the sand this sweet water is found.

September 25.–At sunrise Mr. Hamblin and I part from the train once more, taking with us Chuar, a chief of the Kaibabits, for a trip to the south, for one more view of the Grand Canyon from the summit of the Kaibab Plateau. All day long our way is over red hills, with a bold line of cliffs on our left. A little after noon we reach a great spring, and here we are to camp for the night, for the region beyond us is unknown and we wish to enter it with a good day before us. The Indian goes out to hunt a rabbit for supper, and Hamblin and I climb the cliffs. From an elevation of 1,800 feet above the spring we watch the sun go down and see the sheen on the Vermilion Cliffs and red-lands slowly fade into the gloaming; then we descend to supper.

September 26.–Early in the morning we pass up a beautiful valley to the south and turn westward onto a great promontory, from the summit of which the Grand Canyon is in view. Its deep gorge can be seen to the westward for 50 or 60 miles, and to the southeastward we look off into the stupendous chasm, with its marvelous forms and colors.

Twenty-one years later I read over the notes of that day's experience and the picture of the Grand Canyon from this point is once more before me. I did not know when writing the notes that this was the grandest view that can be obtained of the region from Fremont's Peak to the Gulf of California, but I did realize that the scene before me was awful, sublime, and glorious–awful in profound depths, sublime in massive and strange forms, and glorious in colors. Years later I visited the same spot with my friend Thomas Moran. From this world of wonder he selected a section which was the most interesting to him and painted it. That painting, known as "The Chasm of the Colorado," is in a hall in the Senate wing of the Capitol of the United States. If any one will look upon that picture, and then realize that it was but a small part of the landscape before us on this memorable 26th day of September, he will understand why I suppress my notes descriptive of the scene. The landscape is too vast, too complex, too grand for verbal description.

We sleep another night by the spring on the summit of the Kaibab, and next day we go around to Point Sublime and then push on to the very verge of the Kaibab, where we can overlook the canyon at the mouth of the Little Colorado. The day is a repetition of the glorious day before, and at night we sleep again at the same spring. In the morning we turn to the northeast and descend from Kaibab to the back of Marble Canyon and cross it at the foot of the Vermilion Cliffs, and find our packers camped at Jacob's Pool, where a spring bursts from the cliff at the summit of a great hill of talus. In the camp we find a score or more of Indians, who have joined us here by previous appointment, as we need their services in crossing the river.

On the last day of September we follow the Vermilion Cliffs around to the mouth of the Paria. Here the cliffs present a wall of about 2,000 feet in height,–above, orange and vermilion, but below, chocolate, purple, and gray in alternating bands of rainbow brightness. The cliffs are cut with deep side canyons, and the rainbow hills below are destitute of vegetation. At night we camp on the bank of the Colorado River, on the same spot where our boat-party had camped the year before. Leaving the party in charge of Mr. Graves and Mr. Bishop, while they are building a ferryboat, I take some Indians to explore the canyon of the Paria. We find steep walls on either side, but a rather broad, flat plain below, through which the muddy river winds its way over quicksands. This stream we have to cross from time to time, and we find the quicksands treacherous and our horses floundering in the trembling masses.

These broad canyons, or canyon valleys, are carved by the streams in obedience to an interesting law of corrasion. Where the declivity of the stream is great the river corrades, or cuts its bottom deeper and still deeper, ever forming narrow clefts, but when the stream has cut its channel down until the declivity is greatly reduced, it can no longer carry the load of sand with which it is fed, but drops a part of it on the way. Wherever it drops it in this manner a sand bank is formed. Now the effect of this sand bar is to turn the course of the river against the wall or bank, and as it unloads in one place it cuts in another below and loads itself again; so it unloads itself and forms bars, and loads itself with more material to form bars, and the process of vertical cutting is transformed into a process of lateral cutting. The rate of cutting is greatly increased thereby, but the wear is on the sides and not on the bottom. So long as the declivity of the stream is great, the greater the load of sand carried the greater the rate of vertical cutting; but when the declivity is reduced, so that part of the load is thrown down, vertical cutting is changed to lateral and the rate of corrosion multiplied thereby. Now this broad valley canyon, or "box canyon," as such channels are usually called in the country, has been formed by the stream itself, cutting its channel at first vertically and afterwards laterally, and so a great flood-plain is formed.

For a day we ride up the Paria, and next day return. The party in camp have made good progress. The boat is finished and a part of the camp freight has been transported across the river. The next day the remainder is ferried over and the animals are led across, swimming behind the ferryboat in pairs. Here a bold bluff more than 1,200 feet in height has to be climbed, and the day is spent in getting to its summit. We make a dry camp, that is, without water, except that which has been carried in canteens by the Indians.

October 4--All day long we pass by the foot of the Echo Cliffs, which are in fact the continuation of the Vermilion Cliffs. It is still a landscape of rocks, with cliffs and pinnacles and towers and buttes on the left, and deep chasms running down into the Marble Canyon on the right. At night we camp at a water pocket, a pool in a great limestone rock. We still go south for another half day to a cedar ridge; here we turn westward, climbing the cliffs, which we find to be not the edge of an escarpment with a plateau above, but a long narrow ridge which descends on the eastern side to a level only 500 or 600 feet above the trail left below. On the eastern side of the cliff a great homogeneous sandstone stretches, declining rapidly, and on its sides are carved innumerable basins, which are now filled with pure water, and we call this the Thousand Wells. We have a long afternoon's ride over sand dunes, slowly toiling from mile to mile. We can see a ledge of rocks in the distance, and the Indian with us assures us that we shall find water there. At night we come to the cliff, and under it, in a great cave, we find a lakelet. Sweeter, cooler water never blessed the desert.

While at Jacob's Pool, several days before, I sent a runner forward into this region with instructions to hunt us up some of the natives and bring them to this pool. When we arrive we are disappointed in not finding them on hand, but a little later half a dozen men come in with the Indian messenger. They are surly fellows and seem to be displeased at our coming. Before midnight they leave. Under the circumstances I do not feel that it is safe to linger long at this spot; so I do not lie down to rest, but walk the camp among the guards and see that everything is in readiness to move. About two o'clock I set a couple of men to prepare a hasty lunch, call up all hands, and we saddle, pack, eat our lunch, and start off to the southwest to reach the Moenkopi, where there is a little rancheria of Indians, a farming settlement belonging to the Oraibis, so we are told. We set out at a rapid rate, and when daylight comes we are in sight of the canyon of the Moenkopi, into which we soon descend; but the rancheria has been abandoned. Up the Moenkopi we pass several miles, in a beautiful canyon valley, until we find a pool in a nook of a cliff, where we feel that we can defend ourselves with certainty, and here we camp for the night. The next day we go on to Oraibi, one of the pueblos of the Province of Tusayan.

At Tusayan we stop for two weeks and visit the seven pueblos on the cliffs. Oraibi is first reached, then Shumopavi, Shupaulovi, and Mashongnavi, and finally Walpi, Sichumovi, and Hano.

In a street of Oraibi our little party is gathered. Soon a council is called by the cacique, or chief, and we are assigned to a suite of six or eight rooms for our quarters. We purchase corn of some of the people, and after feeding our animals they are intrusted to two Indian boys, who, under the direction of the cacique, take them to a distant mesa to herd. This is my first view of an inhabited pueblo, though I have seen many ruins from time to time. At first I am a little disappointed in the people. They seem scarcely superior to the Shoshones and Utes, tribes with whom I am so well acquainted. Their dress is less picturesque, and the men have an ugly fashion of banging their hair in front so that it comes down to their eyes and conceals their foreheads. But the women are more neatly dressed and arrange their hair in picturesque coils.

Oraibi is a town of several hundred inhabitants. It stands on a mesa or little plateau 200 or 300 feet above the surrounding plain. The mesa itself has a rather diversified surface. The streets of the town are quite irregular, and in a general way run from north to south. The houses are constructed to face the east. They are of stone laid in mortar, and are usually three or four stories high. The second story stands back upon the first, leaving a terrace over one tier of rooms. The third is set back of the second, and the fourth back of the third; so that their houses are terraced to face the east. These terraces on the top are all flat, and the people usually ascend to the first terrace by a ladder and then by another into the lower rooms. In like manner, ladders or rude stairways are used to reach the upper stories. The climate is very warm and the people live on the tops of their houses. It seems strange to see little naked children climbing the ladders and running over the house tops like herds of monkeys. After we have looked about the town and been gazed upon by the wondering eyes of the men, women, and children, we are at last called to supper. In a large central room we gather and the food is placed before us. A stew of goat's flesh is served in earthen bowls, and each one of us is furnished with a little earthen ladle. The bread is a great novelty to me. It is made of corn meal in sheets as thin and large as foolscap paper. In the corner of the house is a little oven, the top of which is a great flat stone, and the good housewife bakes her bread in this manner: The corn meal is mixed to the consistency of a rather thick gruel, and the woman dips her hand into the mixture and plasters the hot stone with a thin coating of the meal paste. In a minute or two it forms into a thin paper-like cake, and she takes it up by the edge, folds it once, and places it on a basket tray; then another and another sheet of paper-bread is made in like manner and piled on the tray. I notice that the paste stands in a number of different bowls and that she takes from, one bowl and then another in order, and I soon see the effect of this. The corn before being ground is assorted by colors, white, yellow, red, blue, and black, and the sheets of bread, when made, are of the same variety of colors, white, yellow, red, blue, and black. This bread, held on very beautiful trays, is itself a work of art.

They call it piki. After we have partaken of goat stew and bread a course of dumplings, melons, and peaches is served, and this finishes the feast. What seem to be dumplings are composed of a kind of hash of bread and meat, tied up in little balls with cornhusks and served boiling hot. They are eaten with much gusto by the party and highly praised. Some days after we learned how they are made; they are prepared of goat's flesh, bread, and turnips, and kneaded by mastication. As we prefer to masticate our own food, this dainty dish is never again a favorite.

In the evening the people celebrate our advent by a dance, such it seemed to us, but probably it was one of their regular ceremonies.

After dark a pretty little fire is built in the chimney corner and I spend the evening in rehearsing to a group of the leading men the story of my travels in the canyon country. Of our journey down the canyon in boats they have already heard, and they listen with great interest to what I say. My talk with them is in the Mexican patois, which several of them understand, and all that I say is interpreted.

The next morning we are up at daybreak. Soon we hear loud shouts coming from the top of the house. The cacique is calling his people. Then all the people, men, women, and children, come out on the tops of their houses. Just before sunrise they sprinkle water and meal from beautiful grails; then they all stand with bare heads to watch the rising of the sun. When his full orb is seen, once more they sprinkle the sacred water and the sacred meal over the tops of the houses. Then the cacique in a loud voice directs the labor of the day. So his talk is explained to us. Some must gather corn, others must go for wood, water must be brought from the distant wells, and the animals of the strangers must be cared for. Now the house tops present a lively scene. Bowls of water are brought; from them the men fill their mouths and with dexterity blow water over their hands in spray and wash their faces and lave their long shining heads of hair; and the women dress one another's locks. With bowls of water they make suds of the yucca plant, and wash and comb and deftly roll their hair, the elder women in great coils at the back of the head, the younger women in flat coils on their cheeks. And so the days are passed and the weeks go by, and we study the language of the people and record many hundreds of their words and observe their habits and customs and gain some knowledge of their mythology, but above all do we become interested in their religious ceremonies.

One afternoon they take me from Oraibi to Shupaulovi to witness a great religious ceremony. It is the invocation to the gods for rain. We arrive about sundown, and are taken into a large subterranean chamber, into which we descend by a ladder. Soon about a dozen Shamans are gathered with us, and the ceremony continues from sunset to sunrise. It is a series of formal invocations, incantations, and sacrifices, especially of holy meal and holy water. The leader of the Shamans is a great burly bald-headed Indian, which is a remarkable sight, for I have never seen one before. Whatever he says or does is repeated by three others in turn. The paraphernalia of their worship is very interesting. At one end of the chamber is a series of tablets of wood covered with quaint pictures of animals and of corn, and overhead are conventional black clouds from which yellow lightnings are projected, while drops of rain fall on the corn below. Wooden birds, set on pedestals and decorated with plumes, are arranged in various ways. Ears of corn, vases of holy water, and trays of meal make up a part of the paraphernalia of worship. I try to record some of the prayers, but am not very successful, as it is difficult to hold my interpreter to the work. But one of these prayers is something like this:

"Muingwa pash lolomai, Master of the Clouds, we eat no stolen bread; our young men ride not the stolen ass; our food is not stolen from the gardens of our neighbors. Muingwa pash lolomai, we beseech of thee to dip your great sprinkler, made of the feathers of the birds of the heavens, into the lakes of the skies and sprinkle us with sweet rains, that the ground may be prepared in the winter for the corn that grows in the summer."

At one time in the night three women were brought into the kiva. These women had a cincture of cotton about their loins, but were otherwise nude. One was very old, another of middle age, and the third quite young, perhaps fourteen or fifteen years old. As they stood in a corner of the kiva their faces and bodies were painted by the bald-headed priest. For this purpose he filled his mouth with water and pigment and dexterously blew a fine spray over the faces, necks, shoulders, and breasts of the women. Then with his finger as a brush he decorated them over this groundwork, which was of yellow, with many figures in various colors. From that time to daylight the three women remained in the kiva and took part in the ceremony as choristers and dancing performers.

At sunrise we are filed out of the kiva, and a curious sight is presented to our view. Shupaulovi is built in terraces about a central court, or plaza, and in the plaza about fifty men are drawn up in a line facing us. These men are naked except that they wear masks, strange and grotesque, and great flaring headdresses in many colors.

Our party from the kiva stand before this line of men, and the bald-headed priest harangues them in words I cannot understand. Then across the other end of the plaza a line of women is formed, facing the line of men, and at a signal from the old Shaman the drums and the whistles on the terraces, with a great chorus of singers, set up a tumultuous noise, and with slow shuffling steps the line of men and the line of women move toward each other in a curious waving dance. When the lines approach so as to be not more than 10 or 12 feet apart, our party still being between them, they all change so as to dance backward to their original positions. This is repeated until the dancers have passed over the plaza four times. Then there is a wild confusion of dances, the order of which I cannot understand,–if indeed there is any system, except that the men and women dance apart. Soon this is over, and the women all file down the ladder into the kiva and the men strip off their masks and arrange themselves about the plaza, every one according to his own wish, but as if in sharp expectancy; then the women return up the ladder from the kiva and climb to the tops of the houses and stand on the brink of the nearer terrace. Now the music commences once more, and the old woman who was painted in the kiva during the night throws something, I cannot tell what, into the midst of the plaza. With a shout and a scream, every man jumps for it; one seizes it, another takes it away from him, and then another secures it; and with shouts and screams they wrestle and tussle for the charm which the old woman has thrown to them. After a while some one gets permanent possession of the charm and the music ceases. Then another is thrown into the midst. So these contests continue at intervals until high noon.

In the evening we return to Oraibi. And now for two days we employ our time in making a collection of the arts of the people of this town. First, we display to them our stock of goods, composed of knives, needles, awls, scissors, paints, dyestuffs, leather, and various fabrics in gay colors. Then we go around among the people and select the articles of pottery, stone implements, instruments and utensils made of bone, horn, shell, articles of clothing and ornament, baskets, trays, and many other things, and tell the people to bring them the next day to our rooms. A little after sunrise they come in, and we have a busy day of barter. When articles are brought in such as I want, I lay them aside. Then if possible I discover the fancy of the one who brings them, and I put by the articles the goods which I am willing to give in exchange for them. Having thus made an offer, I never deviate from it, but leave it to the option of the other party to take either his own articles or mine lying beside them. The barter is carried on with a hearty good will; the people jest and laugh with us and with one another; all are pleased, and there is nothing to mar this day of pleasure. In the afternoon and evening I make an inventory of our purchases, and the next day is spent in packing them for shipment. Some of the things are heavy, and I engage some Indians to help transport the cargo to Fort Wingate, where we can get army transportation.

October 24-.–To-day we leave Oraibi. We are ready to start in the early morning. The whole town comes to bid us good-by. Before we start they perform some strange ceremony which I cannot understand, but, with invocations to some deity, they sprinkle us, our animals, and our goods with water and with meal. Then there is a time of handshaking and hugging. "Good-by; good-by; good-by!" At last we start. Our way is to Walpi, by a heavy trail over a sand plain, among the dunes. We arrive a little after noon. Walpi, Sichumovi, and Hano are three little towns on one butte, with but little space between them; the stretch from town to town is hardly large enough for a game of ball. The top of the butte is of naked rock, and it rises from 300 to 400 feet above the sand plains below by a precipitous cliff on every side. To reach it from below, it must be climbed by niches and stairways in the rock. It is a good site for defense. At the foot of the cliff and on some terraces the people have built corrals of stone for their asses. All the water used in these three towns is derived from a well nearly a mile away–a deep pit sunk in the sand, over the site of a dune-buried brook.

When we arrive the men of Walpi carry our goods, camp equipage, and saddles up the stairway and deposit them in a little court. Then they assign us eight or ten rooms for our quarters. Our animals are once more consigned to the care of Indian herders, and after they are fed they are sent away to a distance of some miles. There is no tree or shrub growing near the Walpi mesa. It is miles away to where the stunted cedars are found, and the people bring curious little loads of wood on the backs of their donkeys, it being a day's work to bring such a cargo. The people have anticipated our coming, and the wood for our use is piled in the chimney corners. After supper the hours till midnight are passed in rather formal talk.

Walpi seems to be a town of about 150 inhabitants, Sichumovi of less than 100, and Hano of not more than 75. Hano, or "Tewa" as it is sometimes called, has been built lately; that is, it cannot be more than 100 or 200 years old. The other towns are very old; their foundation dates back many centuries–so we gather from this talk. The people of Hano also speak a radically distinct language, belonging to another stock of tribes. They formerly lived on the Rio Grande, but during some war they were driven away and were permitted to build their home here.

Two days are spent in trading with the people, and we pride ourselves on having made a good ethnologic collection. We are especially interested in seeing the men and women spin and weave. In their courtyards they have deep chambers excavated in the rocks. These chambers, which are called kivas, are entered by descending ladders. They are about 18 by 24 feet in size. The kiva is the place of worship, where all their ceremonies are performed, where their cult societies meet to pray for rain and to prepare medicines and charms against fancied and real ailments and to protect themselves by sorcery from the dangers of witchcraft. The kivas are also places for general rendezvous, and at night the men and women bring their work and chat and laugh, and in their rude way make the time merry. Many of the tribes of North America have their cult societies, or "medicine orders," as they are sometimes called, but this institution has been nowhere developed more thoroughly than among the pueblo Indians of this region. I am informed that there are a great number in Tusayan, that a part of their ceremonies are secret and another part public, and that the times of ceremony are also times for feasting and athletic sports.

Here at Walpi the great snake dance is performed. For several days before this festival is held the people with great diligence gather snakes from the rocks and sands of the region round about and bring them to the kiva of one of their clans in great numbers, by scores and hundreds. Most of these snakes are quite harmless, but rattlesnakes abound, and they are also caught, for they play the most important role in the great snake dance. The medicine men, or priest doctors, are very deft in the management of rattlesnakes. When they bring them to the kiva they herd all the snakes in a great mass of writhing, hissing, rattling serpents. For this purpose they have little wands, to the end of each one of which a bunch of feathers is affixed. If a snake attempts to leave its allotted place in the kiva the medicine man brushes it or tickles it with the feather-armed wand, and the snake turns again to commingle with its fellows. After many strange and rather wearisome ceremonies, with dancing and invocations and ululations, the men of the order prepare for the great performance with the snakes. Clothed only in loincloth, each one seizes a snake, and a rattlesnake is preferred if there are enough of them for all. It is managed in this way: The snake is teased with the feather wand and his attention occupied by one man, while another, standing near, at a favorable moment seizes the snake just, back of the head. Then he puts the snake in his mouth, holding it across, so that the head protrudes on one side and the body on the other, which coils about his hand and arm. A few inches of the head and neck are free, and with this free portion the snake struggles, squirming in the air; but the attention of the snake is constantly occupied by the attendant who carries the wand. Then the men of the priest order carrying the snakes in their mouths arrange themselves in a line in the court and move in a procession several times about the court, and then engage in a dance. After the ceremony all of the snakes are carried to the plain and given their freedom.

This snake dance was not witnessed at the time of the first visit, but an account of it was then obtained, such as given above. It has since been witnessed by myself and by others, and carefully prepared accounts of the ceremonies have been published by different persons.

At last our work at Walpi is done, on October 27, and we arrange to leave on the morrow.

CHAPTER XIV. TO ZUNI.

October 28.–To-day we leave the Province of Tusayan for a journey through the Navajo country. There is quite an addition to the party now, for we have a number of Indians employed as freighters. Their asses are loaded with heavy packs of the collections we have made in the various towns of Tusayan. After a while we enter a beautiful canyon coming down from the east, and by noon reach a spring, where we halt for refreshment. The poor little donkeys are thoroughly wearied, but our own animals have had a long rest and have been well fed and are all fresh and active. On the rocks of this canyon picture-writings are etched, and I try to get some account of them from the Indians, but fail.

After lunch we start once more. It is a halcyon day, and with a companion I leave the train and push on for a view of the country. Away we gallop, my Indian companion and I, over the country toward a great plateau which we can see in the distance. The Salahkai is covered with a beautiful forest. We have an exhilarating ride. When the way becomes stony and rough we must walk our horses. My Indian, who is well mounted on a beautiful bay, is a famous rider. About his brow a kerchief is tied, and his long hair rests on his back. He has keen black eyes and a beaked nose; about his neck he wears several dozen strings of beads, made of nacre shining shells, and little tablets of turkis are perforated and strung on sinew cord; in his ears he has silver rings, and his wrists are covered with silver bracelets. His leggings are black velvet, the material for which he has bought from some trader; his moccasins are tan-colored and decorated with silver ornaments, and the trappings of his horse are decorated in like manner. He carries his rifle with as much ease as if it were a cane, and rides with wonderful dexterity. We get on with jargon and sign language pretty well. At night, after a long ride, I descend to the foot of the mesa, and near a little lake I find the camp. The donkey train has not arrived, but soon one after another the Indians come in with their packs, and with white men, Oraibi Indians, Walpi Indians, and Navajos, a good party is assembled.

October 29.–We have a long ride before us to-day, for we must reach old Fort Defiance. I stay with the train in order to keep everything moving, for we expect to travel late in the night. On the way no water is found, but in mid-afternoon the trail leads to the brink of a canyon, and the Indians tell me there is water below; so the animals are unpacked and taken down the cliff in a winding way among the rocks, where they are supplied with water. Again we start; night comes on and we are still in the forest; the trail is good, yet we make slow progress, for some of the animals are weary and we have to wait from time to time for the stragglers. About ten o'clock we descend from the plateau to the canyon beneath and are at old Port Defiance, and the officers at the agency give us a hearty greeting.

We spend the 30th of October at the agency and see thousands of Indians, for they are gathered to receive rations and annuities. It is a wild spectacle; groups of Indians are gambling, there are several horse races, and everywhere there is feasting. At night the revelry is increased; great fires are lighted, and groups of Indians are seen scattered about the plains.

November 1.–After a short day's ride we camp at Rock Spring. A fountain gushes from the foot of the mesa. Then another day's ride through a land of beauty. On the left there is a line of cliffs, like the Vermilion Cliffs of Utah. In the same red sandstones and on the top of the cliff the Kaibab scenery is duplicated. A great tower on the cliff is known as "Navajo Church." Early in the afternoon we are at Fort Wingate and in civilization once more. The fort is on a beautiful site at the foot of the Zuni Plateau. And now our journey with the pack train is ended, and I bid good-by to my Indian friends. My own pack train is to go back to Utah, while from Fort Wingate I expect to go to Santa Fe in an ambulance. But the region about is of interest for its wonderful geologic structure and for the many ruins of ancient pueblos found in the neighborhood. On the 2d of November Captain Johnson, an artillery officer, takes me for a ride among the ruins. Many of these ancient structures are found, but those which are of the most interest are the round towers. Nothing remains of these but the bare walls. They average from 18 to 20 feet in diameter, and are usually two or three stories high.

Probably they were built as places of worship.

Above Fort Wingate there is a great plateau; below, there stretches a vast desert plain with mesas and buttes. The ruins are at the foot of the plateau where the streams come down from the pine-clad heights.

On the 3d of November with a party of officers I visit Zuni in an ambulance. The journey is 40 miles, along the foot of the plateau half the way, and then we turn into the desert valley, in the midst of which runs the Zuni River, sometimes in canyons cut in black lava. Zuni is a town much like those already visited, except that it is a little larger. Nothing can be more repulsive than the appearance of the streets; irregular, crowded, and filthy, in which dogs, asses, and Indians are mingled in confusion. In the distance Toyalone is seen, a great butte on which an extensive ruin is found, the more ancient home of these people, though Zuni itself appears to be hundreds of years old. The people speak a language radically different from that of Tusayan, and no other tribe in the United States has a tongue related to it.

In the midst of the town there is an old Spanish church, partly in ruins, but it is still graced with the wooden image of a saint, gayly colored; and the old tongueless bell remains, for it was sounded with a stone hammer held in the hand of the bellman; the marks of his blows are deeply indented in the metal. Alvar Nunez Caveza de Vaca was the first white man to see Zuni, when he wandered in that long journey from Florida around by the headwaters of the Arkansas, through what is now New Mexico and Arizona, southward to the City of Mexico. He had with him a Barbary negro, who was killed by the Zuni, and his burial place is still pointed out.

Among the Zuni, as among the tribes of Tusayan, the form of government which prevails throughout the North American tribes is well illustrated. Kinship is the tie by which the members of the tribe are bound together as a common body of people. Each tribe is divided into a series of clans, and a clan is a group of people that reckon kinship through the family line. The children therefore belong to the clan of the mother. Marriage is always without the clan; the husband and father must belong to a different clan from the mother and children, and the children belong to their mother and are governed by her brothers, or by her mother's brothers if they be still living. The husband is but the guest of the wife and the clan, and has no other authority in the family than that acquired by personal character. If he is an able and wise man his advice may be taken, but each clan is very jealous of its rights, and the members do not submit to dictation from the guest husband. The woman is not the ruler of the clan; the ruler is the patriarch or elder man, or if he is not a man of ability a younger and more able man is chosen, who by legal fiction is recognized as the elder. Over the officers of the clan are the officers of the tribe,–a chief with assistant chiefs. The organization by tribal governors varies from tribe to tribe. Sometimes the chieftaincy is hereditary in a particular clan, but more often the chieftaincy is elective. There is very little personal property among the tribal people, such property being confined to clothing, ornaments, and a few inconsiderable articles. The ownership of the great bulk of the property inheres in the clan, such as their houses, their patches of land, the food raised from the soil, and the game caught in the chase. Sometimes the clans are grouped, two or more constituting a phratry, and then there are other officers or chiefs standing between the clan and tribal authority. Again, tribes are sometimes organized into confederacies, and a grand confederate chief recognized. In addition to the chieftaincy of confederate tribes, phratries, and clans, there are councils; but these are not councils of legislation in the ordinary sense. The councils are clans whose decisions become a precedent. Tribal law is therefore court-made law, and such customary law grows out of the exigencies which daily life presents to the people. The problems as they arise are solved as best they may be, and the deliberations of the councils look not to the future but only to the present, and are invoked to settle controversy, that peace may be maintained. Of course there is no written constitution or body of laws, but there are traditional regulations which are well preserved in the idioms of oral speech, every rule of procedure or of justice being sooner or later coined into an aphorism.

It has been seen that a clan is a body of kinship in the female line; but the members of the different clans are related to one another by intermarriage. Thus the first tie is by affinity; but, as fathers belong to other clans than the children, the tie is also by consanguinity. Thus the entire tribe is a body of kindred, and the tribal organization is a fabric with warp of streams of blood and woof of marriage ties. When different tribes unite to form a confederacy for offensive or defensive purposes, artificial kinship is established. One tribe perhaps is recognized as the grandfather tribe, another is the father tribe, a third is the elder-brother tribe, a fourth is the younger-brother tribe, etc. In these artificial kinships the members of one tribe address the members of another tribe by kinship terms established in the treaty. Strangers are sometimes adopted into a clan, and this gives them a status in the tribe. The adoption is usually accomplished by the woman claiming the individual as her youngest son or daughter, and such adopted person has thereupon the status belonging to such a natural child; and, though he be an adult, he calls the child born into the clan before his advent, though it be but a year old, his elder brother or his elder sister. Then often young men are advanced in the clan because of superior ability, and this is done by giving them a kinship rank higher than that belonging to their real age; so that it is not infrequently found that old men address young men as their elder brothers and yield to their authority. The ties of the tribe are kinship, and authority inheres in superior age; but in order to adjust these rules so that the abler men may be given control, artificial kinship and artificial age are established. The civil chiefs direct the daily life of the people in their labors.

To the civil organization of the tribe, as thus indicated, there is added a military organization, and war chiefs are selected. But usually these war chiefs are something more than war chiefs, for they also constitute a constabulary to preserve peace and mete out punishment; and young men from the various clans are designated as warriors and advanced in military rank according to merit. There is thus a brotherhood of warriors, and every man in this brotherhood recognizes all others of the group as being elder or younger, and so assumes or yields authority in all matters pertaining to war and the enforcement of criminal law.

In addition to the secular government there is always a cult government. In every tribe there are Shamans, designated variously by white men as "medicine men," "priests," "priest doctors," "theurgists," etc. In many tribes, perhaps in all, the people are organized into Shamanistic societies; but that these societies are invariably recognized is not certain. The Shamans are always found. Among the Zuni there are thirteen of these cult societies. The purpose of Shamanistic institutions is to control the conduct of the members of the tribe in relation to mythic personages, the mysterious beings in which the savage men believe. In the mind of the savage the world is peopled by a host of mythic beings, anthropomorphic and zoomorphic. The difference between man and brute recognized in civilization, is unrecognized in savagery. All animal life is wonderful and magical co sylvan man. Wisdom, cunning, skill, and prowess are attributed to the real animals to a degree often greater than to man; and there are mythic animals as well as mythic men–monsters dwelling in the mountains and caves or hiding in the waters, who make themselves invisible as they pass over the land. Not only are there great monsters, beasts, and reptiles in their mythology, but there are wonderful insects and worms. All life is miraculous and is worshiped as divine. The heavenly bodies, the sun and moon and stars, are mythic animals, and all of the phenomena of nature are attributed to these zoic beings. For example, the Indian knows nothing of the ambient air. The wind is the breath of some beast, or it is a fanning which rises from under the wings of a mythic bird. All the phenomena of nature, the rising and setting of the sun, the waxing and waning of the moon, the shining of the stars, the coming of comets, the flash of meteors, the change of seasons, the gathering and vanishing of the clouds, the blowing of the winds, the falling of the rain, the spreading of the snow, and all other phenomena of physical nature, are held to be the acts of these wonderful zoic deities. It is deemed of prime importance that such deities should be induced to act in the interest of men. Thus it is that Shamanistic government is held to be of as great importance as tribal government, and the Shamans are the peers of the chiefs. With some tribes the cult societies have greater powers than the clan; with other tribes clan government is the more important; but always there is a conflict of authority, and there is a perpetual war between Shamanistic and civil government.

These Shamans and cult societies have a great variety of functions to perform. All disease and all injuries are attributed to mythic beings or to witchcraft, and on these pathologic ideas the medicine practices of the people are based. The medicine men are sorcerers, who work wonders in discovering witchcraft and averting its effects or in discovering the disease-making animals and overcoming their power. So the Shamans and the cult societies are the possessors of medicine and ceremonies designed to prevent and cure human ailments. They also have charge of the ceremonies necessary to avert disaster and to secure success in all the affairs of life in peace and war; and they prescribe methods and observances and furnish charms and amulets, and in every way possible control human conduct in its relation to the unknown. No small part of savage life is devoted to cult ceremonies and observances. The hunter cannot penetrate the forest without his charm; the woman cannot plant corn until a ceremony is performed for securing the blessings of some divine being. Religious festivals and ceremonies are carried on for days and weeks. A war must be submitted to the gods, and a sneeze demands a prayer.

Our arrival at Fort Wingate practically ended the exploration of the great valley of the Colorado. This was in 1870. In 1891 we can look back upon the completion of the survey of all of that region, for it has now been carefully mapped. The geology of the country has been studied, and the tribes which inhabit it have been subjects of careful research. This work has been carried on by a large corps of men, and interesting results have accrued.

CHAPTER XV. THE GRAND CANYON.

The Grand Canyon is a gorge 217 miles in length, through which flows a great river with many storm-born tributaries. It has a winding way, as rivers are wont to have. Its banks are vast structures of adamant, piled up in forms rarely seen in the mountains.

Down by the river the walls are composed of black gneiss, slates, and schists, all greatly implicated and traversed by dikes of granite. Let this formation be called the black gneiss. It is usually about 800 feet in thickness.

Then over the black gneiss are found 800 feet of quartzites, usually in very thin beds of many colors, but exceedingly hard, and ringing under the hammer like phonolite. These beds are dipping and unconformable with the rocks above; while they make but 800 feet of the wall or less, they have a geological thickness of 12,000 feet. Set up a row of books aslant; it is 10 inches from the shelf to the top of the line of books, but there may be 3 feet of the books measured directly through the leaves. So these quartzites are aslant, and though of great geologic thickness, they make but 800 feet of the wall. Your books may have many-colored bindings and differ greatly in their contents; so these quartzites vary greatly from place to place along the wall, and in many places they entirely disappear. Let us call this formation the variegated quartzite.

Above the quartzites there are 500 feet of sandstones. They are of a greenish hue, but are mottled with spots of brown and black by iron stains. They usually stand in a bold cliff, weathered in alcoves. Let this formation be called the cliff sandstone.

Above the cliff sandstone there are 700 feet of bedded sandstones and limestones, which are massive sometimes and sometimes broken into thin strata. These rocks are often weathered in deep alcoves. Let this formation be called the alcove sandstone.

Over the alcove sandstone there are 1,600 feet of limestone, in many places a beautiful marble, as in Marble Canyon. As it appears along the Grand Canyon it is always stained a brilliant red, for immediately over it there are thin seams of iron, and the storms have painted these limestones with pigments from above. Altogether this is the red-wall group. It is chiefly limestone. Let it be called the red wall limestone.

Above the red wall there are 800 feet of gray and bright red sandstone, alternating in beds that look like vast ribbons of landscape. Let it be called the banded sandstone.

And over all, at the top of the wall, is the Aubrey limestone, 1,000 feet in thickness. This Aubrey has much gypsum in it, great beds of alabaster that are pure white in comparison with the great body of limestone below. In the same limestone there are enormous beds of chert, agates, and carnelians. This limestone is especially remarkable for its pinnacles and towers. Let it be called the tower limestone.

Now recapitulate: The black gneiss below, 800 feet in thickness; the variegated quartzite, 800 feet in thickness; the cliff sandstone, 500 feet in thickness; the alcove sandstone, 700 feet in thickness; the red wall limestone, 1,600 feet in thickness; the banded sandstone, 800 feet in thickness; the tower limestone, 1,000 feet in thickness.

These are the elements with which the walls are constructed, from black buttress below to alabaster tower above. All of these elements weather in different forms and are painted in different colors, so that the wall presents a highly complex facade. A wall of homogeneous granite, like that in the Yosemite, is but a naked wall, whether it be 1,000 or 5,000 feet high. Hundreds and thousands of feet mean nothing to the eye when they stand in a meaningless front. A mountain covered by pure snow 10,000 feet high has but little more effect on the imagination than a mountain of snow 1,000 feet high–it is but more of the same thing; but a facade of seven systems of rock has its sublimity multiplied sevenfold.

Let the effect of this multiplied facade be more clearly realized.

Stand by the river side at some point where only the black gneiss is seen. A precipitous wall of mountain rises over the river, with crag and pinnacle and cliff in black and brown, and through it runs an angular pattern of red and gray dikes of granite. It is but a mountain cliff which may be repeated in many parts of the world, except that it is singularly naked of vegetation, and the few plants that find footing are of strange tropical varieties and are conspicuous because of their infrequency.

Now climb 800 feet and a point of view is reached where the variegated quartzites are seen. At the summit of the black gneiss a terrace is found, and, set back of this terrace, walls of elaborate sculpture appear, 800 feet in height. This is due to the fact that though the rocks are exceedingly hard they are in very thin layers or strata, and these strata are not horizontal, but stand sometimes on edge, sometimes highly inclined, and sometimes gently inclined. In these variegated beds there are many deep recesses and sharp salients, everywhere set with crags, and the wall is buttressed by a steep talus in many places. In the sheen of the midday sun, these rocks, which are besprinkled with quartz crystals, gleam like walls of diamonds.

A climb of 800 feet over the variegated beds and the foot of the cliff sandstone is reached. It is usually olive green, with spots of brown and black, and presents 500 feet of vertical wall over the variegated sandstone. The dark green is in fine contrast with the variegated beds below and the red wall above.

Climb these 500 feet and you stand on the cliff sandstone. A terrace appears, and sometimes a wall of terraces set with alcoves of marvelous structure. Climb to the summit of this alcove sandstone–700 feet–and you stand at the foot of the red wall limestone. Sometimes this stands in two, three, or four Cyclopean steps–a mighty stairway. Oftener the red wall stands in a vertical cliff 1,600 feet high. It is the most conspicuous feature of the grand facade and imparts its chief characteristic. All below is but a foundation for it; all above, but an entablature and sky-line of gable, tower, pinnacle, and spire. It is not a plain, unbroken wall, but is broken into vast amphitheaters, often miles abound, between great angular salients.

The amphitheaters also are broken into great niches that are sometimes vast chambers and sometimes royal arches 500 or 1,000 feet in height.

Over the red wall limestone, with its amphitheaters, chambers, niches, and royal arches–a climb of 1,600 feet–is the banded sandstone, the entablature over the niched and columned marble, an adamantine molding 800 feet in thickness, stretching along the walls of the canyon through hundreds of miles. This banded sandstone has massive strata separated by friable shales. The massive strata are the horizontal elements in the entablature, but the intervening shales are carved with a beautiful fretwork of vertical forms, the sculpture of the rills. The massive sandstones are white, gray, blue, and purple, but the shales are a brilliant red; thus variously colored bands of massive rock are separated by bands of vertically carved shales of a brilliant hue.

On these highly colored beds the tower limestone is found, 1,000 feet in height. Everywhere this is carved into towers, minarets, and domes, gray and cold, golden and warm, alabaster and pure, in wonderful variety.

Such are the vertical elements of which the Grand Canyon facade is composed. Its horizontal elements must next be considered. The river meanders in great curves, which are themselves broken into curves of smaller magnitude. The streams that head far back in the plateau on either side come down in gorges and break the wall into sections. Each lateral canyon has a secondary system of laterals, and the secondary canyons are broken by tertiary canyons; so the crags are forever branching, like the limbs of an oak. That which has been described as a wall is such only in its grand effect. In detail it is a series of structures separated by a ramification of canyons, each having its own walls. Thus, in passing down the canyon it seems to be inclosed by walls, but oftener by salients–towering structures that stand between canyons that run back into the plateau. Sometimes gorges of the second or third order have met before reaching the brink of the Grand Canyon, and then great salients are cut off from the wall and stand out as buttes–huge pavilions in the architecture of the canyon. The scenic elements thus described are fused and combined in very different ways.

We measured the length of the Grand Canyon by the length of the river running through it, but the running extent of wall cannot be measured in this manner. In the black gneiss, which is at the bottom, the wall may stand above the river for a few hundred yards or a mile or two; then, to follow the foot of the wall, you must pass into a lateral canyon for a long distance, perhaps miles, and then back again on the other side of the lateral canyon; then along by the river until another lateral canyon is reached, which must be headed in the black gneiss. So, for a dozen miles of river through the gneiss, there may be a hundred miles of wall on either side. Climbing to the summit of the black gneiss and following the wall in the variegated quartzite, it is found to be stretched out to a still greater length, for it is cut with more lateral gorges. In like manner, there is yet greater length of the mottled, or alcove, sandstone wall; and the red wall is still farther stretched out in ever branching gorges. To make the distance for ten miles along the river by walking along the top of the red wall, it would be necessary to travel several hundred miles. The length of the wall reaches its maximum in the banded sandstone, which is terraced more than any of the other formations. The tower limestone wall is less tortuous. To start at the head of the Grand Canyon on one of the terraces of the banded sandstone and follow it to the foot of the Grand Canyon, which by river is a distance of 217 miles, it would be necessary to travel many thousand miles by the winding Way; that is, the banded wall is many thousand miles in length.

Stand at some point on the brink of the Grand Canyon where you can overlook the river, and the details of the structure, the vast labyrinth of gorges of which it is composed, are scarcely noticed; the elements are lost in the grand effect, and a broad, deep, flaring gorge of many colors is seen. But stand down among these gorges and the landscape seems to be composed of huge vertical elements of wonderful form. Above, it is an open, sunny gorge; below, it is deep and gloomy. Above, it is a chasm; below, it is a stairway from gloom to heaven.

The traveler in the region of mountains sees vast masses piled up in gentle declivities to the clouds. To see mountains in this way is to appreciate the masses of which they are composed. But the climber among the glaciers sees the elements of which this mass is composed,–that it is made of cliffs and towers and pinnacles, with intervening gorges, and the smooth billows of granite seen from afar are transformed into cliffs and caves and towers and minarets. These two aspects of mountain scenery have been seized by painters, and in their art two classes of mountains are represented: mountains with towering forms that seem ready to topple in the first storm, and mountains in masses that seem to frown defiance at the tempests. Both classes have told the truth. The two aspects are sometimes caught by our painters severally; sometimes they are combined. Church paints a mountain like a kingdom of glory. Bierstadt paints a mountain cliff where an eagle is lost from sight ere he reaches the summit. Thomas Moran marries these great characteristics, and in his infinite masses cliffs of immeasurable height are seen.

Thus the elements of the facade of the Grand Canyon change vertically and horizontally. The details of structure can be seen only at close view, but grand effects of structure can be witnessed in great panoramic scenes. Seen in detail, gorges and precipices appear; seen at a distance, in comprehensive views, vast massive structures are presented. The traveler on the brink looks from afar and is overwhelmed with the sublimity of massive forms; the traveler among the gorges stands in the presence of awful mysteries, profound, solemn, and gloomy.

For 8 or 10 miles below the mouth of the Little Colorado, the river is in the variegated quartzites, and a wonderful fretwork of forms and colors, peculiar to this rock, stretches back for miles to a labyrinth of the red wall cliff; then below, the black gneiss is entered and soon has reached an altitude of 800 feet and sometimes more than 1,000 feet; and upon this black gneiss all the other structures in their wonderful colors are lifted. These continue for about 70 miles, when the black gneiss below is lost, for the walls are dropped down by the West Kaibab Fault, and the river flows in the quartzites.

Then for 80 miles the mottled, or alcove, sandstones are found in the river bed. The course of the canyon is a little south of west and is comparatively straight. At the top of the red wall limestone there is a broad terrace, two or three miles in width, composed of hills of wonderful forms carved in the banded beds, and back of this is seen a cliff in the tower limestone. Along the lower course of this stretch the whole character of the canyon is changed by another set of complicating conditions. We have now reached a region of volcanic activity. After the canyons were cut nearly to their present depth, lavas poured out and volcanoes were built on the walls of the canyon, but not in the canyon itself, though at places rivers of molten rock rolled down the walls into the Colorado.

The next 80 miles of the canyon is a compound of that found where the river is in the black gneiss and that found where the dead volcanoes stand on the brink of the wall. In the first stretch, where the gneiss is at the foundation, we have a great bend to the south, and in the last stretch, where the gneiss is below and the dead volcanoes above, another great southern detour is found. These two great beds are separated by 80 miles of comparatively straight river. Let us call this first great bend the Kaibab reach of the canyon, and the straight part the Kanab reach, for the Kanab Creek heads far off in the plateau to the north and joins the Colorado at the beginning of the middle stretch. The third great southern bend is the Shiwits stretch. Thus there are three distinct portions of the Grand Canyon of the Colorado: the Kaibab section, characterized more by its buttes and salients; the Kanab section, characterized by its comparatively straight walls with volcanoes on the brink; and the Shiwits section, which is broken into great terraces with gneiss at the bottom and volcanoes at the top.

The Grand Canyon of the Colorado is a canyon composed of many canyons. It is a composite of thousands, of tens of thousands, of gorges. In like manner, each wall of the canyon is a composite structure, a wall composed of many walls, but never a repetition. Every one of these almost innumerable gorges is a world of beauty in itself. In the Grand Canyon there are thousands of gorges like that below Niagara Palls, and there are a thousand Yosemites. Yet all these canyons unite to form one grand canyon, the most sublime spectacle on the earth. Pluck up Mt. Washington by the roots to the level of the sea and drop it headfirst into the Grand Canyon, and the dam will not force its waters over the walls. Pluck up the Blue Ridge and hurl it into the Grand Canyon, and it will not fill it.

The carving of the Grand Canyon is the work of rains and rivers. The vast labyrinth of canyon by which the plateau region drained by the Colorado is dissected is also the work of waters. Every river has excavated its own gorge and every creek has excavated its gorge. When a shower comes in this land, the rills carve canyons–but a little at each storm; and though storms are far apart and the heavens above are cloudless for most of the days of the year, still, years are plenty in the ages, and an intermittent rill called to life by a shower can do much work in centuries of centuries.

The erosion represented in the canyons, although vast, is but a small part of the great erosion of the region, for between the cliffs blocks have been carried away far superior in magnitude to those necessary to fill the canyons. Probably there is no portion of the whole region from which there have not been more than a thousand feet degraded, and there are districts from which more than 30,000 feet of rock have been carried away. Altogether, there is a district of country more than 200,000 square miles in extent from which on the average more than 6,000 feet have been eroded. Consider a rock 200,000 square miles in extent and a mile in thickness, against which the clouds have hurled their storms and beat it into sands and the rills have carried the sands into the creeks and the creeks have carried them into the rivers and the Colorado has carried them into the sea. We think of the mountains as forming clouds about their brows, but the clouds have formed the mountains. Great continental blocks are upheaved from beneath the sea by internal geologic forces that fashion the earth. Then the wandering clouds, the tempest-bearing clouds, the rainbow-decked clouds, with mighty power and with wonderful skill, carve out valleys and canyons and fashion hills and cliffs and mountains. The clouds are the artists sublime.

In winter some of the characteristics of the Grand Canyon are emphasized. The black gneiss below, the variegated quartzite, and the green or alcove sandstone form the foundation for the mighty red wall. The banded sandstone entablature is crowned by the tower limestone. In winter this is covered with snow. Seen from below, these changing elements seem to graduate into the heavens, and no plane of demarcation between wall and blue firmament can be seen. The heavens constitute a portion of the facade and mount into a vast dome from wall to wall, spanning the Grand Canyon with empyrean blue. So the earth and the heavens are blended in one vast structure.

When the clouds play in the canyon, as they often do in the rainy season, another set of effects is produced. Clouds creep out of canyons and wind into other canyons. The heavens seem to be alive, not moving as move the heavens over a plain, in one direction with the wind, but following the multiplied courses of these gorges. In this manner the little clouds seem to be individualized, to have wills and souls of their own, and to be going on diverse errands–a vast assemblage of self-willed clouds, faring here and there, intent upon purposes hidden in their own breasts. In the imagination the clouds belong to the sky, and when they are in the canyon the skies come down into the gorges and cling to the cliffs and lift them up to immeasurable heights, for the sky must still be far away. Thus they lend infinity to the walls.

The wonders of the Grand Canyon cannot be adequately represented in symbols of speech, nor by speech itself. The resources of the graphic art are taxed beyond their powers in attempting to portray its features. Language and illustration combined must fail. The elements that unite to make the Grand Canyon the most sublime spectacle in nature are multifarious and exceedingly diverse. The Cyclopean forms which result from the sculpture of tempests through ages too long for man to compute, are wrought into endless details, to describe which would be a task equal in magnitude to that of describing the stars of the heavens or the multitudinous beauties of the forest with its traceries of foliage presented by oak and pine and poplar, by beech and linden and hawthorn, by tulip and lily and rose, by fern and moss and lichen. Besides the elements of form, there are elements of color, for here the colors of the heavens are rivaled by the colors of the rocks. The rainbow is not more replete with hues. But form and color do not exhaust all the divine qualities of the Grand Canyon. It is the land of music. The river thunders in perpetual roar, swelling in floods of music when the storm gods play upon the rocks and fading away in soft and low murmurs when the infinite blue of heaven is unveiled. With the melody of the great tide rising and falling, swelling and vanishing forever, other melodies are heard in the gorges of the lateral canyons, while the waters plunge in the rapids among the rocks or leap in great cataracts. Thus the Grand Canyon, is a land of song. Mountains of music swell in the rivers, hills of music billow in the creeks, and meadows of music murmur in the rills that ripple over the rocks. Altogether it is a symphony of multitudinous melodies. All this is the music of waters. The adamant foundations of the earth have been wrought into a sublime harp, upon which the clouds of the heavens play with mighty tempests or with gentle showers.

The glories and the beauties of form, color, and sound unite in the Grand Canyon–forms unrivaled even by the mountains, colors that vie with sunsets, and sounds that span the diapason from tempest to tinkling raindrop, from cataract to bubbling fountain. But more: it is a vast district of country. Were it a valley plain it would make a state. It can be seen only in parts from hour to hour and from day to day and from week to week and from month to month. A year scarcely suffices to see it all. It has infinite variety, and no part is ever duplicated. Its colors, though many and complex at any instant, change with the ascending and declining sun; lights and shadows appear and vanish with the passing clouds, and the changing seasons mark their passage in changing colors. You cannot see the Grand Canyon in one view, as if it were a changeless spectacle from which a curtain might be lifted, but to see it you have to toil from month to month through its labyrinths. It is a region more difficult to traverse than the Alps or the Himalayas, but if strength and courage are sufficient for the task, by a year's toil a concept of sublimity can be obtained never again to be equaled on the hither side of Paradise.

Made in the USA
Lexington, KY
11 November 2013